DRY PORTS – A GLOBAL PERSPECTIVE

Dry Ports – A Global Perspective

Challenges and Developments in Serving Hinterlands

Edited by

RICKARD BERGQVIST
Gothenburg University, Sweden

GORDON WILMSMEIER
Economic Commission for Latin America and the Caribbean (ECLAC)

KEVIN CULLINANE
Edinburgh Napier University, UK

Routledge
Taylor & Francis Group

LONDON AND NEW YORK

First published 2013 by Ashgate Publishing

2 Park Square, Milton Park, Abingdon, Oxon OX14 4RN
711 Third Avenue, New York, NY 10017, USA

Routledge is an imprint of the Taylor & Francis Group, an informa business

First issued in paperback 2016

British Library Cataloguing in Publication Data
Dry ports : a global perspective. -- (Transport and society)
  1. Freight and freightage--Management. 2. Marine terminals. 3. Infrastructure (Economics)
  I. Series II. Bergqvist, Rickard. III. Wilmsmeier, Gordon.
  IV. Cullinane, Kevin.
  387.1'64-dc23

Library of Congress Cataloging-in-Publication Data
Bergqvist, Rickard.
Dry ports - a global perspective : challenges and developments in serving hinterlands / by Rickard Bergqvist, Gordon Wilmsmeier, and Kevin Cullinane.
  p. cm. -- (Transport and society)
  Includes bibliographical references and index.
  ISBN 978-1-4094-4424-4 (hbk)
  1. Container terminals. 2. Harbors. 3. Shipping. 4. Freight and freightage. I. Wilmsmeier, Gordon. II. Cullinane, Kevin. III. Title.
  HE551.B37 2013
  388'.044--dc23

ISBN 978-1-4094-4424-4 (hbk)   2012029574
ISBN 978-1-138-27446-4 (pbk)

Contents

List of Figures

List of Tables

List of Contributors

Rickard Bergqvist, PhD, is assistant professor in Logistics and Transport Management at the Transport and Logistics Research Group, School of Business, Economics and Law, University of Gothenburg. His key research areas are maritime logistics, regional logistics, intermodal transportation, dry ports and public–private collaboration. His major works include over 20 refereed journal articles, conference papers and book chapters related to intermodal transport, dry ports, economic modelling and public–private collaboration. His work has appeared in the *Journal of Transport Geography*, *Maritime Economics and Logistics*, *Transport Reviews*, *World Review of Intermodal Transportation Research*, *Journal of Interdisciplinary Economics*, *Transportation Planning and Technology*, *International Journal of Logistics*, *Research and Applications*, *Supply Chain Forum*, *Place Branding and Public Diplomacy* and *Mapping and Image Science*.

Jeroen Bozuwa joined Ecorys (formerly NEI) after his study Transportation and Logistics (with specializations in operations research, freight transport research, materials management, physical distribution and business management) at the Polytechnics of Tilburg. At Ecorys he currently holds the position of Principal Consultant. He has wide experience in the field of freight transport and logistics, with emphasis on the inland transport modes and intermodal transport. His key experience is in the field of research into macroeconomic effects of trends and policy measures related to freight transport, especially freight transport by road, inland waterways and rail. Moreover, he has been involved in (inter)national corridor studies, modal shift studies, studies into efficiency improvements in freight transport, freight logistics studies, impact assessments of policy measures and evaluation studies. Jeroen has professional experience of 20 years.

Zheng Chang is a PhD student in Transportation Management College at Dalian Maritime University. She got her Bachelor of Engineering at Dalian Maritime University in 2008, majoring in Transportation (Ports Management and Administration). She obtained her Engineering Master's with a major in Transportation Planning and Management in 2010, focusing on research on port development and maritime economics. She published "Research on development of container sea-rail intermodal transportation in China" in 2010 and "VAR model analysis on Panamax shipping market's fluctuation" in 2009. She has participated in eight research projects, such as development planning for the Shandong and Liaoning provinces; research on strategic development of the Dalian and Shandong Peninsula Northeast Asian international shipping centre; research on container transport system in Shenzhen; research on the relation between rational economy and the port of Shenzhen; research on the resource allocation of Port Dalian (PDA) Co., Ltd. during the process of regionalization, etc.

Kevin Cullinane is Director of the Transport Research Institute and Professor of International Logistics at Edinburgh Napier University. He was formerly Chair in Marine Transport and Management at Newcastle University, Professor and Head of the Department of Shipping and Transport Logistics at the Hong Kong Polytechnic University, Head of the Centre for International Shipping and Transport at Plymouth University, Senior Partner in his own transport consultancy company and Research Fellow at the University of Oxford Transport Studies Unit. He is a Fellow of the Chartered Institute of Logistics and Transport and Member of the Nautical Institute. He has been a logistics adviser to the World Bank and transport adviser to the governments of Scotland, Ireland, Hong Kong, Egypt, Chile, Korea and the UK. He holds an Honorary Professorship at the University of Hong Kong and is a Visiting Professor in International Logistics at the University of Gothenburg. He has published nine books and over 160 refereed journal and conference papers and is currently Project Head of the EU-funded Northern Maritime University. Kevin is an Associate Editor of *Transportation Research A: Policy and Practice* and the *International Journal of Applied Logistics.* He also sits on the Editorial Boards of *Maritime Economics and Logistics, Transport Reviews*, the *International Journal of Logistics Management*, the *Journal of Logistics and Sustainable Transport*, the *Annals of Maritime Studies*, the *Journal of Shipping and Logistics* and the *Proceedings of IMechE Part M: Journal of Engineering for the Maritime Environment*. Kevin's expertise has recently been recognized at UK national level with his appointment to the Civil and Construction Engineering sub-panel for REF 2014.

Raghu Dayal A former founding Managing Director of Container Corporation of India (CONCOR), a Government of India undertaking for the development of intermodal infrastructure in the country, and currently a Senior Fellow at the Asian Institute of Transport Development, dealing with integrated development and multimodal logistics, Raghu Dayal has had extensive exposure to operational and commercial aspects of transport planning and operations, integrated multimodal logistics. Senior policy-making assignments for promotion of countries' international trade and close interaction with international agencies such as UNECAFE/UNESCAP and UNCTAD enabled him to acquire valuable experiences in trade facilitation and infrastructure, as well as planning for investments across different sectors of the economy and their integration. He was in charge of India's participation in World Expo at Montreal (Canada) as well as Expo in Osaka (Japan) and set up the Indo-German Export Promotion project, funded by the governments of Germany and India, that helped to facilitate trade and investment between India and the European Union (especially Germany). He has been on Boards of Directors for a number of companies in India, and is a regular contributor to some of the leading newspapers and periodicals.

Salvador Furió is Director of Logistics at the Valenciaport Foundation. He is an Industrial Engineer from the Polytechnic University of Valencia, MoS in Ports Management and Intermodal Transport from the Comillas University (ICADE), and holds a Diploma of Advanced Studies from the PhD program in Advanced Models for Operations Management and Supply Chain Management at the Polytechnic University of Valencia. He has participated in and directed many research, consultancy and cooperation projects at national, European and international level. These projects all related to container logistics, maritime, railway and intermodal transport, trade facilitation, the planning and design of logistics platforms and to energy efficiency and environmental management. He has regularly taught and collaborated with different Master's programs of the Universities of Valencia and Castellon and the Polytechnic Universities of Valencia and Barcelona, and he has participated in national and international congresses on transport and logistics.

Johan Gille started working at NEI (nowadays Ecorys) in 2000 after completing his Master's thesis at TNO on the cluster economics of inland shipping. He holds an MSc in Science and Policy from Utrecht University. Within Ecorys Johan was a project manager in the field of maritime and inland freight transport, with a focus on competition and sustainability. Johan often assesses feasibility of new transport services, mode shift or the construction of new infrastructure or equipment. Moreover he is leading projects dealing with greening sectors and raising sustainability levels throughout transport chains.

Dr Charles Kunaka is a Senior Trade Specialist with the International Trade Department of the World Bank. He works mostly on trade facilitation and logistics in developing countries, where he is involved in projects to improve the performance of international trade corridors and logistics services. A significant proportion of his work is in landlocked countries, which often face higher transport costs than their coastal neighbours. More recently, Charles has been developing measures to improve domestic logistics services in lagging regions within developing countries. Such regions are often uncompetitive in regional and global markets and suffer from the interplay between low economic density and poor logistics services. Before joining the World Bank Charles spent several years as a senior officer at the Southern African Development Community (SADC). While at SADC he led measures to harmonize transport policies across Southern Africa, so as to create an integrated logistics market. Charles has published several papers on transport in Sub-Saharan Africa.

Bruce Lambert is the Executive Director at the Institute for Trade and Transportation Studies (ITTS). ITTS is a multistate research institution formed to assist member states on understanding the relationship of infrastructure planning and policy to changes in international and domestic freight traffic. Previously, Mr Lambert served as a Senior Economist at the US Army Corps of Engineers, Institute of Water Resources. His work focused on developing economic data and tools to examine the role of Corps investments on port facilities. He also served as the Secretary to the US Section of the International Navigation Association (PIANC). While serving as PIANC Secretary, Mr Lambert provided technical assistance to Latin American ports and countries on matters related to transportation improvements. Prior to joining the USACE, Mr Lambert worked at the Federal Highway Administration. While at FHWA, Mr Lambert researched the nature of freight movements to support national freight policy and planning research. Mr Lambert managed the Freight Analysis Framework study; the first large-scale project to map and outline the underlying transportation flows of the United States economy for national and regional policy research. Mr Lambert also worked at the Port of Long Beach, Standard and Poor's DRI (now IHS Global Insight), and Louisiana State University.

Erick Leal Matamala started his studies on maritime and port economics at the Institute of Transport and Maritime Management ITMMA, University of Antwerp, Belgium, where obtained an MSc in Transport and Maritime Economics during 2007. Currently, he is working on his PhD thesis, developing a model whose main interest is revealing the role of capabilities and knowledge on port development. Other related fields of interest are logistic platforms and industrial organization with focus on the West Coast South American Port Industry. All this work is being carried out in close collaboration with United Nations Economic Commission for Latin America and the Caribbean (UN-ECLAC), namely the Infrastructure Service Unit. Mr Leal's main affiliation is with the Catholic University of Concepción, Chile.

Jing Lu is a professor, PhD supervisor and Dean of Transportation Management College at Dalian Maritime University. He got his Master's degree at Dongbei University of Finance and Economics and did research on maritime and logistics management at Cardiff University as a visiting scholar. His research fields include maritime planning and management; maritime economics and finance; maritime technology economics and projects assessment and management. He has published more than 120 papers in domestic and overseas journals or conferences, nine of which are included in *EI*. He is the author of *Engineering Economics*, *International Shipping Economics and Market* and *Research on Shandong Peninsula Northeast Asia International Shipping Center*. He directed or participated in more than 40 research projects for governments or enterprises. He was granted the First Class Award for Excellent Research Achievement by the Liaoning Development and Reform Commission in 2008, the May Day Labour Award by Liaoning in 2004, the Municipal Labour Model Award of Dalian in 2004, and the First Class Award for Excellent Teachers by Dalian City in 2003. He is a member of IAME, the Eastern Asia Society for Transportation Studies, the China Communications and Transportation Association, the China University Association of Sea Transportation Education and the Dalian Logistics Association.

Chad R. Miller, PhD, is an Assistant Professor in the University of Southern Mississippi Department of Economic and Workforce Development, and Assistant Director of the Center for Logistics, Trade, and Transportation.

Jason Monios is a Research Fellow with the Maritime Research Group at TRI (the Transport Research Institute, Edinburgh Napier University, UK). His primary research area is intermodal freight transport, with a specific interest in the relations between ports and inland terminals in managing hinterland access. A particular focus of his work is on the roles of government policy and planning, and he is currently exploring the potential of institutional approaches in understanding these connections. In the last two years he has conducted interviews at ports and intermodal terminals throughout Europe and the United States, the results of which are forthcoming in a number of journal papers and book chapters. Other recent projects have included research into the viability of short sea shipping in Scotland and a two-month consultancy post at UNCTAD. He has a Master's degree in Transport Planning and Engineering from Edinburgh Napier University, and he is currently undertaking a PhD in Transport Geography, researching freight hub development in Scotland. Jason also holds a PhD in English Literature from the University of Sydney, Australia.

Libby Ogard, President and Principal of Prime Focus LLC, spent 17 years with Burlington Northern Railroad and Conrail and several years at Schneider National Carriers, where she was Retail Division General Manager. Prime Focus LLC was established in 2001 as a freight transportation consulting firm specializing in economic analysis, freight transportation research, freight policy issues, transportation facility feasibility studies and public outreach. She is actively involved with the Transportation Research Forum and the Transportation Research Board, among other professional organizations.

Gabriel Pérez Salas Gabriel Pérez-Salas is Civil Engineer with a Masters in Maritime and Port Management. With more than 14 years of experience in the United Nations System, nowadays is Associate Economic Affairs Officer in the Infrastructure Services Unit at the United Nations Economic Commission for Latin America and the Caribbean (UN-ECLAC). His main areas of interest are logistics, dry ports, surface transport and intelligent transport systems (ITS).

Ben J. Ritchey has 30 years of transportation consulting related experience, primarily for the US Departments of Transportation (US DOT), Homeland Security (US DHS), and Energy (US DOE), as well as in the transportation industry. His professional experience ranges from a General Manager with profit/loss responsibilities, business development with sales and strategy responsibilities, and a President of a small non-profit research organization. He also has Congressional experience. He has served as Vice President/General Manager of Battelle's Transportation Division with a revenue of approximately $70 million/year. Currently he is President of his firm, Acadia Group, LLC. Mr Ritchey has managed a multimillion-dollar indefinite delivery/indefinite quantity technical support contract with the Federal Highway Administration (FHWA), US DOT. He has managed a number of visible projects for the FHWA including Strategic Multimodal Analysis, Multimodal Freight Analysis Framework, US DOT Truck Size and Weight, Federal Highway Cost Allocation and several Ohio-based PPP projects (i.e. Heartland Corridor). He also led several US DOE projects including alternative fuels and hazardous material routing. Mr Ritchey's transportation policy experience includes serving as a former Chair of the Transportation Research Board's (TRB) Freight Data Committee and Co-Chair of the Columbus Regional Logistics Council, which serves the Columbus regional freight community. He also sits on three advisory boards: The University of Michigan Trucking Advisory Board, the North Carolina A&T Transit Advisory Board, and Franklin University Supply Chain Advisory Board.

Dr Leo Tadeu Robles, Brazilian, 63 years old, is a graduate in Economics Sciences (1971), a Master in Administration (1995) and has a PhD in Administration (2001) from São Paulo University. He is an invited professor on postgraduate courses in São Paulo (SP) and São Luis (MA). At Santos Catholic University he coordinated the International Logistics and Maritime Economics Studies Group (NELIEMA), teaching in the postgraduate program in Business Management and graduate courses. He has professional experience in Private and Public Administration, focused in Transportation and Logistics, as well as Planning and Transport Projects Evaluation. His research interests focus on International Logistics, Maritime Economics, Business Administration, Environmental Management and Foreign Trade issues.

Gavin Roser is Secretary General of the European Freight and Logistics Leaders Forum which brings together global shippers and transport providers to review best practice across the supply chain. Gavin is also the Managing Director of Pantrak Transportation Limited – a project management and consultancy firm dealing with all aspects of marine transport, rail and road freight on a pan-European basis. He is also Deputy Chairman and co-founder of the Coastlink Shipping Network and a Director of TruckTrain Knowhow Ltd, a joint venture developing new rail solutions for the carriage of freight. Partner with nine European Universities in the Northern Maritime University a 3-year European Interreg IVB programme. Previous Board appointments include the Denholm Shipping Group of Companies and Chairman of a joint venture in the Falklands. He has worked on overseas assignments in Iran and New Zealand, the latter involving the privatization of the National Shipping Line. He has been in management for European Rail businesses (CP Ships) and General Management positions with subsidiary of the BAA (subsequently acquired by Securicor). He is a former Vice Consul for Norway in Glasgow. Mr Roser has had extensive speaking engagements at European Transport Forums, and most recently in May 2012 EU Green Week in Brussels on "maximizing the commercial potential of the Danube" and a major shipping event in Odessa with the theme – "maximizing container shipping opportunities in the Black Sea". Gavin is a member of Tactran Transport Partnership, appointed by Minister of Transport – Scotland from 2007 to the present.

Kenneth Russell has over 27 years of transport experience within the Russell Group. Having come through the ranks of the transport business, Kenneth has a comprehensive knowledge of warehousing, distribution, rail freight operations and logistics. Director for the development and marketing since 1995, Kenneth has been instrumental in the design and structure of major supply chain logistic contracts managed by a Scottish-based, privately-owned company which currently employs in excess of 500 staff.

Ricardo J. Sánchez is an economist working at the United Nations Economic Commission for Latin America and the Caribbean (UN-ECLAC). He is currently chief of the Infrastructure Services Unit, in charge of maritime, logistic and port issues as well as transport infrastructure and regulatory matters. He is an economist and has an MSc in Administration and Economics of Public Utilities at the University of Paris X, France, and University Carlos III of Madrid, Spain. Ricardo J. Sánchez is an internationally recognized expert in shipping and port economics, with special focus in Latin-American and Caribbean cases. He has developed his professional career in more than 35 countries since 1983, mostly related to transport and infrastructure. His main research interests are shipping and port economics, including the maritime cycle, port devolution, national maritime policies and industrial organization applied to shipping markets. He has more than 100 publications in the form of books, chapters in books, peer-reviewed articles and working papers. At the same time he is a member of the Council of the International Association of Maritime Economists (IAME), and member of the Port Performance Research Network (PPRN) and the Argentine Association of Economists.

Vaibhav Shah is a chartered member of Chartered Institute of Logistics and Transport (CILT), India. He has completed his postgraduate from Shipping and Transport College, Rotterdam, the Netherlands and also from the Institute of Rail Transport in India. He has over 11 years of experience with India's first and major dry port operator – CONCOR – and has built up a wealth of experience as specialist in the dry port sector. Terminal Planning and Development, Contract Management, Operations, Tariff setting, Traffic Survey and feasibility study are among his areas of expertise. He has broad experience in various aspects of the Container and Intermodal Transport Industry. He is involved in the planning of new dry ports, expansion, development of dry port terminals in Gujarat and is also involved in the planning of Gujarat's first Multimodal Freight Logistics Park by CONCOR, with much of his work using innovative and global thinking to produce optimum output of facilities. He has visited various ports and logistics centers in the Netherlands, Germany, Belgium and England, and has also visited France, Italy and Scotland. He contributes to educating tomorrow's Port and Logistics Managers at Gujarat University since 2006 and has also visited a few more institutes, such as Ahmadabad Management Association, CEPT University, H.L. College for Professional Education (HLCPE) and Future Innoversity, working in Container and Intermodal Transport, Container Traffic Control, Port Development, Terminal Management, Supply Chain Management.

Gordon Wilmsmeier, PhD, is Economic Affairs Officer at the United Nations Economic Commission for Latin America and the Caribbean (UN-ECLAC). Previously, he worked as Principal Research Fellow at the Transport Research Institute (TRI) at Edinburgh Napier University. He holds a PhD in Economic Geography from Osnabruck University, Germany. Gordon was involved in the Interreg IVb North Sea Region projects: NMU (project leader), Dry Port, STRATMoS, FoodPort, LoPinod and MTC. He has also led the 7th Framework Programme project KnowME and has collaborated in various other 4th, 5th, 6th and 7th framework projects. He is an internationally recognized expert in maritime transport and port development issues and has worked as consultant to UN-ECLAC, UNCTAD, UNOHRLLS, The World Bank, Andean Development Fund (CAF), Inter-American Development Bank (IDB), UNDP, Organization of American States (OAS), IIRSA, ProInversion, Peru, JICA and the Ministry of Economy in El Salvador. Gordon has worked on projects in 20 of the 33 countries in Latin America and the Caribbean over the last nine years. Since 2010, Gordon has been an invited lecturer at Goeteborg University, Sweden, Bremen University of Applied Sciences, Germany and the Instituto de Telecomunicaciones, Transporte y Puertos (ITTP), Bogota, Colombia. His over 30 refereed articles were published in: *Journal of Transport Geography, Maritime Economics and Logistics, Transport Reviews, Transportation Research A: Policy and Practice, Maritime Policy and Management, Zeitschrift fuer Wirtschaftsgeographie* among others he also published various book chapters as well as numerous reports and working papers. Gordon is a member of the Council of the International Association of Maritime Economists (IAME) since 2010 and an IAME member since 2002, as well as a member of the Port Performance Research Network (PPRN) and PortEconomics.eu.

Introduction:
A Global Perspective on Dry Ports

Rickard Bergqvist, Gordon Wilmsmeier and Kevin Cullinane

Background

The importance of logistics increases as the economy becomes more and more specialised and globalised. Changes in business environments such as globalisation, production patterns, urbanisation and environmental awareness further support this trend. Since production and logistics arrive at a consensus where every individual product or module is produced in regions where the comparative advantage is the greatest, there is an increased focus on hinterlands and logistics. Traditionally, ports have been in the focus as logistic centres of maritime logistics chains, but changes in production patterns are supported by the development of the rapid transport of goods over long distances. As a result, the relevance of port hinterland transport, high utilisation of transport resources and infrastructure through the consolidation of goods flows and extending the influence of ports in their hinterlands to increase their competitiveness has become even more important. This development emphasises the connection between the intra-regional transport systems and the larger inter-regional transport systems, since this is where much of the consolidation of freight flow occurs.

From an environmental perspective, it is untenable to await direct solutions based on significant technological breakthroughs in the field of alternative energy sources or in increased engine performance. Therefore, other more indirect measures are useful for improving transportation systems. The increased utilisation of transportation resources, the coordination and consolidation of goods flows and increased use of less environmentally damaging means of transport and intermodal solutions are examples of such indirect measures which rest upon the logic of collaboration in a regional setting.

Global container trade and, in particular, container ports are facing challenges related to capacity expansion, environmental considerations, and community restrictions. At the same time, freight transport and logistics functions are more and more integrated into global supply chains. The challenges for the container trade and liner shipping have moved inland from the sea, first to the ports and then to the hinterland (cf. Notteboom 2002). The increased scale of ships puts more pressure on ports as they must handle large volumes of load units during short periods of time (Cullinane and Khanna 1999). Being able to effectively and efficiently distribute the load units to and from the hinterland is crucial for overall efficiency at the ports and, ultimately, for the whole supply chain (Cullinane and Khanna

2000). As a consequence, costs and lead time are increasingly being generated in the smaller routes, rather than in the arteries (Bergqvist and Woxenius 2011).

The use of high capacity transport modes, such as trains and barges, is one measure to increase the capacity of hinterland transport. Both rail and inland waterway present some advantages in terms of decreased environmental impact, economies of scale, faster throughput in ports and less delay related to road congestion. Maximising hinterland effectiveness and efficiency is a matter of finding the optimal mix of transport modes and setups, rather than identifying a single service or solution.

Improving the hinterland connectivity of ports has become more and more important for addressing today's logistics challenges. The hinterlands of ports have been able to expand due to containerisation in combination with intermodal transport possibilities (Song 2003). As hinterlands expand, the hinterlands of different ports naturally overlap and inter-port competition intensifies (cf. Notteboom and Winkelmans 2001, Cullinane and Wilmsmeier 2011). This intensified competition, in combination with the complexity of hinterland transport and associated infrastructure and strategic transhipment nodes, have made hinterland connectivity an essential part of a port's distinct value proposition (Bergqvist 2011). The potential for more effective and efficient hinterland systems, associated with better collaboration and coordination in the supply chain, gives hinterland logistics and associated concepts, such as dry ports, an obvious role in designing and managing global supply chains.

The development worldwide concerning "dry ports" (in their various forms, functions and strategies) addresses many of the challenges facing contemporary logistics and ports. The concept of a dry port is more often used in practice while being given more scientific attention. In 1982, the UN first used the term to underline the integration of services with different traffic modes under one contract (Beresford and Dubey 1990). A "dry port" was defined as an inland terminal to and from which shipping lines could issue their bills of lading (UNCTAD 1982). The concept has evolved from merely focusing on the container segment to other market segments as well, so that it is now more focussed on the services originally offered at the port but moved inland (Woxenius and Bergqvist 2011, Cullinane and Wilmsmeier 2011). Parallel to the development of the concept in practice and theory, numerous definitions have been developed (e.g., Rodrigue et al. 2010, Van den Bossche and Gujar 2010, Cardebring and Warnecke 1995, Ng and Gujar 2009, UNESCAP 2006, Roso et al. 2009, Jaržemskis and Vasiliauskas 2007, Harrison et al. 2002, Leitner and Harrison 2001, Walter and Poist 2003). Although alternative definitions do exist, there seems to be a consensus on the importance of dry ports; potential dry ports must improve cost-efficiency, environmental performance (e.g. congestion, pollution, safety, health, and noise) and the logistics quality of hinterland logistics (cf. Bergqvist and Woxenius 2011, Roso et al. 2009, Padilha and Ng 2011).

This book comprises a number of case studies and state-of-the-art examples from measures taken in different parts of the world with varying economic, social, institutional and environmental realities that show the complexity and diverse approaches to this phenomenon.

Dry Ports: A Global Phenomena with Local Characteristics

The contributions in this book illustrate dry port applications in Europe, Africa, Asia, and North America.

In Chapter 2, Rickard Bergqvist describes the development of hinterland transport in Sweden and in particular the system of rail shuttles connected to the Port of Gothenburg. This development has been possible due to a number of reasons, deregulation being the most evident enabler. The development of new intermodal terminals has been a prerequisite for the system of rail shuttles to expand geographically and in this chapter a number of key factors related to the development of dry ports are identified and described. The research concludes that the existence of commitment and trust is crucial since the development process is facing stress from many different sources.

The contribution by Bergqvist (2013) describes the remarkable journey within some sub-segments of hinterland logistics and transport in Europe related to the development of dry ports and freight villages.

There is an increasing interest in the concept of dry ports from policy makers and logistics service providers; the case study by Gille and Bozuwa (2013), presented in Chapter 3, provides an example of how potential dry port establishments can be evaluated, analysed and assessed. The case study recognises the role of dry ports as economic drivers within the regions in which they are located, by focussing on the potential of a dry port development in the Southeast Drenthe region of the Netherlands by defining three questions:

1. Are sufficient volumes of freight being transported between main seaports and the region, or passing along the region, to allow for multimodal freight services concentrated in a dry port?
2. Are the required infrastructure and services in place?
3. How do stakeholders envisage the potential of a dry port in this region?

These questions illustrate the necessary components involved in dry port development, i.e. material flows, infrastructure and stakeholders. Issues of coordination, freight volumes and bundling are important aspects highlighted by stakeholders. Furthermore, the regulatory and operational framework and the behaviour and perspectives of stakeholders are identified as key development factors (cf. Bergqvist 2013).

Connectivity regarding the relationship between the hinterland and the seaport is a multifaceted issue. The concept relates to both the physical transportation with associated extended services of the port, as well as the virtual connection related to communication and information exchange. The virtual connection is often not given priority; however, much research concludes that this is an area with great inefficiency and potential for improvement (cf. Furió 2013). In Chapter 4, Salvador Furió builds on the integration of maritime and rail operations. Special attention is given to information integration and the role of Port Community Systems. Furió develops a standard for computerising information exchange between stakeholders and a pilot test involving the Madrid Dry port and the Port of Valencia was carried out. The analysis identifies great potential and the pilots confirmed significant benefits associated with a common standard for information exchange. Benefits include service quality improvement and cost reductions (operational and administrative costs) in maritime-rail operations in dry ports and seaport terminals.

In Chapter 5, Gavin Roser, Kenneth Russel, Gordon Wilmsmeier and Jason Monios emphasise the importance of a user perspective when planning infrastructure development. The chapter provides insights from one of Scotland's most innovative logistics service providers on the planning, location and utilisation of hinterland intermodal terminals. The lack of low wagons in the UK is identified as a large barrier for future rail-based intermodal transport growth. Current regulations and government funding do not provide a solution to this problem and the lead-time can be detrimental to the service development process. Better utilisation of resources and better information for potential shippers can help in improving the competitiveness and attractiveness of train services. In sum, major adverse impacts have been observed from the current arrangements. The hinterland transport requirements of Scottish trade flows have not been given the necessary attention, either in policy or by the private sector, where a visionary mid- and long-term perspective is not common.

Benefits associated with dry ports and related intermodal transport services are often referred to within the context of large efficient hubs with modern and high-capacity infrastructure. However, the benefits are often the greatest in areas and markets with high trade/transaction costs. This is especially true for regions such as Africa which also have a large number of landlocked countries (cf. Kunaka 2013). In Chapter 6, Charles Kunaka addresses the issue of high logistics and trade costs and the role that dry ports may have in decreasing these trade barriers. The author observes that previously, road, rail and port projects were often designed and developed in isolation, rather than being designed as an integrated part of a transport system. Today, more strategic and coherent approaches to logistics are being adopted in many regions where port development is becoming more linked to hinterland transport systems. For many regions in Africa, this should mean more efficient access to global markets and should also facilitate intraregional trade (cf. Kunaka 2013).

In Chapter 7, Ragdu Dayal describes the development of intermodal transport in India by Indian Railways and its subsidiary Container Corporation of India (CONCOR). An extensive countryside network of ICDs (inland container depots) and CFSs (container freight stations), along with a comprehensive institutional framework, have all been developed. Although a somewhat late-starter in developing the multimodal infrastructure, India has taken great strides in its steady and sustainable growth, with a comprehensive framework of systems, procedures and institutions already in place. The development and operation of dry ports in the country provides new challenges to face. These challenges involve the further penetration of hinterlands, along with the need to consolidate and coalesce facilities which have mushroomed in some areas, simultaneously with the need to expeditiously redress some practices which appear to distort the sector and have the potential to debilitate the system.

In Chapter 8, Vaibhav Shah investigates the dynamics of price, cost and quality of Indian dry ports. Due to improper planning, fiscal constraints, differences in status, non-standardisation, and imbalanced port–dry port integration, the author identifies the need for government intervention to bring standardisation, better quality of services and overall development of the sector. Once the standards are achieved, operations will become more harmonised and best practices more easily identified. The author asserts that better facilitation and a stringent regulatory environment would support future growth, with the goal being to maintain a steady and relatively competitive situation, with less focus on low pricing and more focus on quality and efficiency at Indian dry ports.

In Chapter 9, Jing Lu and Zheng Chang introduce the status of dry port development in China. Economic strategies which focus on a vast inland with tremendous resources and great potential have triggered an enthusiasm for dry port construction on the part of coastal ports in China. Port authorities have noticed that conventional port competition has evolved into a competition between the supply chains in which ports are involved. For seaports, a dry port can bring a port's function forward in these supply chains, to an inland city for example, and can provide efficient hinterland transport and access. It may also ease a port's demand for land for expansion, by moving logistics activities to the city. At the same time, for inland regions, the diverse functions that a dry port possesses may attract more investment and promote the local economy. These are seen as great opportunities to coordinate and balance Chinese economic development.

The empirical analysis uses Dalian as an example to introduce the characteristics, modes of construction, and the cooperation strategies taken by coastal ports and dry ports in the actual operational process in China. From the empirical analysis, the authors suggest that port operators should have closer cooperation with local government and should try to strengthen the relationship amongst other stakeholders such as shipping companies, customs, and railway operators.

In Chapter 10, Bruce Lambert, Chad Miller, Libby Ogard and Ben Ritchey provide a discussion of the role of dry ports in the United States, largely framing the role of dry ports as one element in a broader transportation network. Roles associated with the linkages between ports and hinterlands in the United States, with a specific emphasis on railway linkages, are also presented. The chapter ends with a discussion of the institutional challenges facing dry ports and opportunities related to dry port development.

In Chapter 11, Jason Monios and Bruce Lambert compare port hinterland access strategies in the form of intermodal freight corridors connecting ports and inland intermodal terminals. Detailed case studies of the Alameda Corridor, the Alameda Corridor East, Norfolk Southern's Heartland Corridor and CSX's National Gateway are presented in this chapter. The various projects represent corridors of different sizes, objectives and challenges relating to stakeholder management. The results indicate the importance of aligning stakeholder objectives with funding sources and planning schedules. Of particular importance to the development of hinterland access is recent US policy towards the provision of public funding through discretionary funding programmes. These developments are discussed in the context of US transport policy and the difficulties of government involvement in a traditionally privately owned and operated rail industry.

In Chapter 12, Leo Tadeus Robles describes the development process of dedicated areas for foreign trade in Brazil and, in particular, the Santos Metropolitan region. From a Brazilian regulatory and policy perspective, foreign trade is facilitated and enhanced in two dimensions. The first dimension concerns logistics, focusing on location sites, infrastructure and overall connectivity. The second dimension is based on legal regulations that are not necessarily related to a specific site or location. The author concludes that it is important that the various roles of dedicated areas are shared by Governmental Agencies in a vision for enhancing foreign trade as part of a more general objective towards poverty reduction, creating jobs and wealth generation opportunities open to all people.

In Chapter 13, Erick Leal Matamala, Gabriel Pérez Salas and Ricardo J. Sánchez examine the potential for developing logistics zones in Chile. By analysing the decisions of logistics operators regarding location selection, the authors are able to identify the potential for developing logistics zones. The methodology is based on an econometric model that contains variables influencing the location decision process. Combined with a cluster analysis, 49 Chilean provinces are classified. The results show that Santiago, Valparaíso and Concepción are the provinces with the highest potential, with market characteristics and port infrastructure as the main influences over this potential.

In fast growing economic regions, as described in Chapters 7–9 and 12–13, dry ports have evidently become a necessity for relieving the ever-increasing stress experienced by seaports in dealing with rapid economic expansion (e.g., India – Dayal 2013, Shah 2013; China – Lu and Chang 2013; Brazil – Robles 2013; Chile – Leal Matamala et al. 2013). Limited space at container yards, stress on the road infrastructure and congestion at the gates of the seaport are examples of effects that increase the need for better hinterland connectivity and the development of dry ports.

Conclusions

The benefits associated with dry ports and intermodal transport usually fall within the categories of cost-efficiency, environmental performance and logistics quality

(Bergqvist; 2013). As the benefits are enjoyed by many stakeholders, the interest in the concept of dry ports is multifaceted. Public actors and decision-makers often associate the dry port concept with improved competitiveness of local and regional businesses, increased attractiveness of the region and sustainable logistics development (Bergqvist 2008). As a result, the development of dry ports is often a process involving public actors, often in partnership with private actors, i.e. public–private partnerships (Bergqvist 2008). The involvement of public actors raises some interesting issues related to the institutional framework. Tendering, concession agreements, independency of terminal operations, transparency, ownership, responsibilities and roles are examples of difficult aspects that need to be considered in every dry port development process (Bergqvist 2013). Case studies suggest that these issues look quite similar regionally (e.g. in Europe) as well as globally, but are addressed in many different ways (cf. Shah 2011, Bergqvist 2013, Roser et al. 2013).

Public–private partnerships have the potential to balance the development process of dry ports by accounting for objectives and benefits from both a private and public perspective. The partnership may with great advantage utilise the characteristics and benefits of the different actors, e.g. the private actors' closeness to the market and the public actors' access to infrastructure investment funds and long-term perspectives, to name a few. The issue of aligning the different time-perspectives of different actors and aligning them with the funding sources and planning schedules is an important part of a successful development process (Monios and Lambert 2013, Bergqvist 2008).

In order to cope with these challenges and aspects, a number of key factors have been identified, e.g. local enthusiasm, formal arrangements and operational framework (Bergqvist 2013), PPP, information exchange (Furió 2013), port connectivity (Kunaka 2013), aligning stakeholder objectives with funding sources and planning schedules (Monios and Lambert 2013).

The concept of dry ports is continuously developing and new generations of dry ports continue to emerge, with an increasing number of sophisticated services being offered (Kunaka 2013, Gille and Bozuwa 2013, Bergqvist 2013). The increasing number of dry ports and the interest in collaboration and exploring the market opportunities for intermodal transport have led to an increasing interest in, and pressure on, the interconnection between nodes (i.e. particularly with respect to railways) and overall transport efficiency and capacity (Kunaka 2013, Bergqvist 2013).

The dry port concept has already generated great benefits globally and has been proven to provide logistics efficiency, low environmental impact and high logistics quality. Regional and local transport policies and infrastructure plans need to consider and address the contemporary challenges associated with the development of intermodal transport in general and dry ports in particular. On this basis, the best incentives for developing effective, efficient and sustainable transport systems can be identified. However, within this context, it is important to recognise that dry ports are a global phenomenon with local characteristics.

Acknowledgements

The authors would like to express their gratitude to the Interreg IVb North Sea Region Programme who financed, in part, the Dry Port Conference in Edinburgh and the book as part of the Dry Port Project (cf. www.dryport.org).

References

Beresford, A.K.C. and Dubey, R.C. 1990. *Handbook on the Management and Operation of Dry Ports*. Geneva: UNCTAD.

Bergqvist, R. 2008. Realizing Logistics Opportunities in a Public–Private Collaborative Setting: The Story of Skaraborg. *Transport Reviews*, 28(2), 219–237.

Bergqvist, R. 2011. Hinterland Logistics and Global Supply Chains, in *Maritime Logistics: Logistics Management of Shipping and Ports*, edited by D.-W. Song and P. Panayides. London: Kogan Page, forthcoming.

Bergqvist, R. 2013. Hinterland Transport in Sweden: The Context of Intermodal Terminals and Dry Ports, in *Dry Ports: A Global Perspective*, edited by R. Bergqvist, G. Wilmsmeier and K.P.B Cullinane. Farnham: Ashgate Publishing Limited.

Bergqvist, R. and Woxenius, J. 2011. The Development of Hinterland Transport by Rail: The Story of Scandinavia and the Port of Gothenburg. *Journal of Interdisciplinary Economics*, 23(2), 161–177.

Bossche, M. van den and Gujar, G. 2010. Competition, Excess Capacity and Pricing of Dry Ports in India: Some Policy Implications. *Journal of Shipping and Transport Logistics*, 2(2), 151–167.

Cardebring, P.W. and Warnecke, C. 1995. *Combi-terminal and Intermodal Freight Centre Development*. KFB-Swedish Transport and Communication Research Board, Stockholm.

Cullinane, K.P.B. and Khanna, M. 1999. Economies of Scale in Large Container Ships. *Journal of Transport Economics and Policy*, 33(2), 185–208.

Cullinane, K.P.B. and Khanna, M. 2000. Economies of Scale in Large Containerships: Optimal Size and Geographical Implications. *Journal of Transport Geography*, 8(3), 181–195.

Cullinane, K.P.B. and Wilmsmeier, G. 2011. The Contribution of the Dry Port Concept to the Extension of Port Life Cycles, in *Handbook of Terminal Planning*, edited by J.W. Böse. Operations Research Computer Science Interfaces Series, Vol. 49. Heidelberg: Springer, 359–380.

Dayal, R. 2013. Dry Port: The India Experience and What the Future Holds – India Needs to Think Out-of-the-Box, in *Dry Ports: A Global Perspective*, edited by R. Bergqvist, G. Wilmsmeier and K.P.B Cullinane. Farnham: Ashgate Publishing Limited.

Furió, S. 2013. Port Community Systems in Maritime and Rail Transport Integration: The Case of Valencia, Spain, in *Dry Ports: A Global Perspective*, edited by R. Bergqvist, G. Wilmsmeier and K.P.B Cullinane. Farnham: Ashgate Publishing Limited.

Gille, J. and Bozuwa, J. 2013. Dry Ports: A Concept or a Reality for Southeast Drenthe?, in *Dry Ports: A Global Perspective*, edited by R. Bergqvist, G. Wilmsmeier and K.P.B Cullinane. Farnham: Ashgate Publishing Limited.

Harrison, R., McCray, J.P., Henk, R. and Prozzi, J. 2002. *Inland Port Transportation: Evaluation Guide*. Center for Transportation Research, University of Texas at Austin, USA.

Jaržemskis, A. and Vasiliauskas, A.V. 2007. Research on Dry Port Concept as Intermodal Node, *Transport*, 22(3), 207–213.

Kunaka, C. 2013. Dry Ports and Trade Logistics in Africa, in *Dry Ports: A Global Perspective*, edited by R. Bergqvist, G. Wilmsmeier and K.P.B Cullinane. Farnham: Ashgate Publishing Limited.

Leal Matamala, E., Pérez Salas, G. and Sánchez, R.J. 2013. Potential for Logistics Zones Development: Chile as a Case Study, in *Dry Ports: A Global Perspective*, edited by R. Bergqvist, G. Wilmsmeier and K.P.B Cullinane. Farnham: Ashgate Publishing Limited.

Leitner, S. J. and Harrison, R. 2001. *The Identification and Classification of Inland Ports*. Research Report Number 0-4083-1, Center for Transportation Research, University of Texas at Austin.

Lu, J. and Chang, Z. 2013. The Construction of Seamless Supply Chain Networks: The Development of Dry Ports in China, in *Dry Ports: A Global Perspective*, edited by R. Bergqvist, G. Wilmsmeier and K.P.B Cullinane. Farnham: Ashgate Publishing Limited.

Monios, J. and Lambert, B. 2013. Intermodal Freight Corridor Development in the United States, in *Dry Ports: A Global Perspective*, edited by R. Bergqvist, G. Wilmsmeier and K.P.B Cullinane. Farnham: Ashgate Publishing Limited.

Ng, K.Y.A and Gujar, G.C. 2009. The Spatial Characteristics of Inland Transport Hubs: Evidences from Southern India, *Journal of Transport Geography*, 17(5), 346–356.

Notteboom, T. and W. Winkelmans. 2001. Structural Changes in Logistics: How will Port Authorities Face the Challenge?, *Maritime Economics and Logistics,* 28(1), 71–89.

Notteboom, T. 2002. The Interdependence between Liner Shipping Networks and Intermodal Networks. *IAME Panama 2002 Conference Proceedings*, Panama City, 13–15 November.

Padilha, P. and Ng, A.K.Y. 2011. The Spatial Evolution of Dry Ports in Developing Economies: The Brazilian Experience. *Maritime Economics and Logistics*, edited by K.P.B Cullinane, R. Bergqvist and G. Wilmsmeier, Special Issue on Dry ports, forthcoming.

Robles, L.T. 2013. Implementing Dedicated Areas for Foreign Trade in the Santos Metropolitan Region: The Brazilian Experience, in *Dry Ports: A Global Perspective*, edited by R. Bergqvist, G. Wilmsmeier and K.P.B Cullinane. Farnham: Ashgate Publishing Limited.

Rodrigue, J.P., Debrie, J., Fremont, A. and Gouvernal, E. 2010. Functions and Actors of Inland Ports: European and North American Dynamics. *Journal of Transport Geography*, 18(4), 519–529.

Roser, G., Russel, K., Wilmsmeier, G. and Monios, J. 2013. Integrating Ports and Hinterlands: A Scottish Perspective from the Shop Floor, in *Dry Ports: A Global Perspective*, edited by R. Bergqvist, G. Wilmsmeier and K.P.B Cullinane. Farnham: Ashgate Publishing Limited.

Roso, V., Woxenius, J. and Lumsden, K. 2009. The Dry Port Concept: Connecting Seaports with their Hinterland by Rail, *Journal of Transport Geography*, 17(5), 338–345.

Shah, V. 2013. Price versus Quality or Quality versus Price at Indian Dry Ports: Cost, Quality and Price – A Visionary View on Indian Dry Ports, in *Dry Ports: A Global Perspective*, edited by R. Bergqvist, G. Wilmsmeier and K.P.B Cullinane. Farnham: Ashgate Publishing Limited.

Song, D.-W. 2003. Port Co-operation in Concept and Practice. *Maritime Policy and Management*, 30(1), 29–44.

UNCTAD. 1982. *Multimodal Transport and Containerisation*, TD/B/C.4/238/ Supplement 1, Part 5: Ports and Container Depots.

UNESCAP. 2006. Logistics Sector Developments: Planning Models for Enterprises and Logistics Clusters. Economic and Social Commission for Asia and the Pacific.

Walter, C.K. and Poist, R.F. 2003. Desired Attributes of an Inland Port: Shipper vs Carrier Perspectives. *Transportation Journal*, 42(5), 42–55.

Woxenius, J. and Bergqvist, R. 2011. Hinterland Transport by Rail: Comparing the Scandinavian Conditions for Maritime Containers and Semi-trailers. *Journal of Transport Geography*, forthcoming issue.

PART I
Europe

Chapter 2

Hinterland Transport in Sweden: The Context of Intermodal Terminals and Dry Ports

Rickard Bergqvist

2.1   Introduction

Because of the environmental impact of heavy road transport and the absence of direct technical solutions, it is of the utmost importance that alternative and indirect solutions are identified. One such possibility is the transfer of freight to more sustainable modes of transport. Intermodal freight transport, building on the connection of two or more modes where high-capacity transport modes are used for the majority of the journey is one enabler of model shift. Intermodal transport has enjoyed a growing interest from academia, industry, and the public sector in general and policy-makers in particular. Parallel to improved cost-efficiency, growing market segments, and increased general interest in intermodal transport, more focus has been put on the development of inland terminals for handling and transshipment of load units.

It is not an easy or quick process to design and establish intermodal terminals. As this chapter illustrates, developing intermodal terminals is especially complex if regional governments, such as municipalities, manage the process.

This chapter aims to increase knowledge and understanding of the development process associated with dry ports and intermodal terminals in the context of Swedish conditions and hinterland logistics. Furthermore, this chapter tries to identify key factors and actors in the development process of intermodal terminals and describe how these interact and affect the process. The chapter also aims to identify and describe the trends and development related to the hinterland road–rail intermodal transport segment.

2.2   The Development of Road–Rail Intermodal Transport in Sweden

The volume of the intermodal transport market in Sweden has been fairly steady until 2000, when the development of the Scandinavian port shuttles system started and volumes increased (see Figure 2.1).

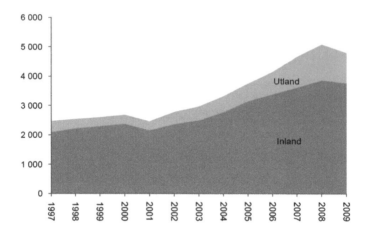

Figure 2.1   Transport performance by rail-based intermodal transport in million tkm.

Source: Trafikanalys 2010

Figure 2.1 indicates a substantial increase in both traffic with domestic destinations (Domestic) and cross-border traffic (International). The Scandinavian rail shuttle system related to Port of Gothenburg is defined as "Domestic" traffic.

2.3   Hinterland Transport in Sweden

The development of dry ports and associated rail shuttles in Scandinavia has been remarkable during the last decade. The central components in the system are the Port of Gothenburg and 26 hinterland rail shuttles to 23 different destinations and dry ports in Scandinavia. Some eleven different rail operators exist in the system (Port of Gothenburg 2011b), an impressive number given that the rail sector in Sweden started its deregulation in 1988. Each shuttle has a frequency of at least three departures per week in each direction. The most frequent one supports the retailer HandM's central warehouse in Eskilstuna, and operates about 14 times a week in each direction.

Most shuttles serve distant dry ports of 200 or more kilometres and conform to traditional hinterland transport. However, there are an increasing amount of shuttles serving much shorter distances, traditionally operated by road. The shortest shuttle runs a distance of about ten kilometres within the city of Gothenburg. In general, independent terminal operators manage the inland terminals, especially the large ones, while local logistics service providers generally operate the small terminals.

The system of shuttles and dry ports handled about 360,000 20-foot equivalent units (TEUs) in 2009 and has a turnover of about €60 million annually (Bergqvist 2009). The container rail shuttle system handled about 40 per cent of all containers to and from the Port of Gothenburg.

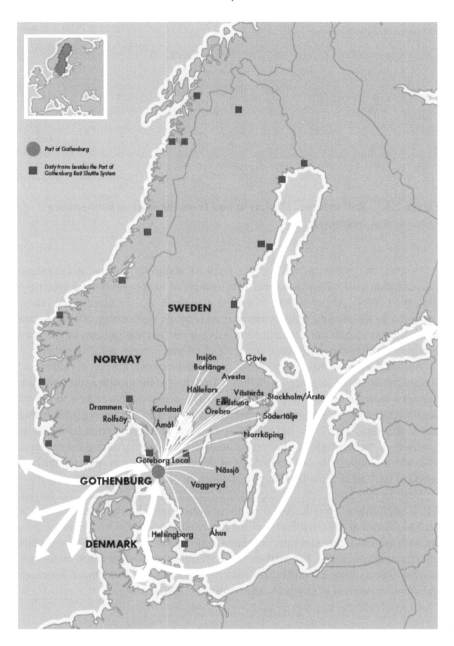

Figure 2.2   The Port of Gothenburg rail shuttle system as of February 2011
Source: Port of Gothenburg 2011b

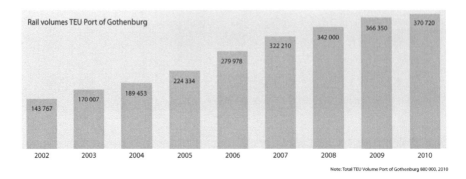

Figure 2.3   Rail volumes (TEUs) to and from the Port of Gothenburg
Source: Port of Gothenburg 2011b

In total, the system of rail shuttles employs about 400 people and decreases transportation cost by up to ten per cent compared to the direct road alternative (Bergqvist 2009). About 51,000t of carbon dioxide (CO_2) are saved every year as compared to the amount used by direct road (Port of Gothenburg 2011a). Recent developments include the introduction of a five-level "grading" system of the inland terminals that illustrates the scope and scale of offered services at the terminals.

Around the time of the millennium, the board of directors at the Port of Gothenburg decided to implement a strategy with the goal that half of the growth in the hinterland container segment should be carried by rail. The achievements related to the rail shuttle system have exceeded the expectations and goals and have gained market shares at a rapid pace. The rail shuttle system has achieved a remarkable annual growth over the last eight to nine years and displayed stability during the economic recession in 2009–2010. However, most of the development has occurred during a period of substantial growth in the general segment of container liner shipping.

Despite its impressive development, there are numerous factors and barriers that hinder future growth. Some of these factors can be described as critical success factors at different levels of the hinterland transportation system, e.g., at the port, the links, the terminals or related to framework issues such as regulation and planning processes. The next section describes the most common factors and in what aspects they are critical for hinterland transport development in general and the development of dry ports and terminals in particular.

2.4   Key Factors Related to Intermodal Terminal Development

This section is based on case studies and research related to the hinterland transport system in Scandinavia (e.g. Bergqvist 2009; Bergqvist 2008b; Bergqvist, Falkemark et al. 2008; Bergqvist, Falkemark et al. 2007; Bergqvist 2008a; Bergqvist 2007; Roso 2006). The section starts out by analyzing the issue of necessary market potential for the establishment of intermodal terminals and rail-based hinterland transport services.

2.4.1   *Market Potential*

The market base and profitability of potential intermodal transport services are obviously the most important factors for developing inland terminals and associated traffic. They usually also speed up the development process by creating an intense pressure on decision-makers, and at the same time it is easier to secure necessary investments and funds. The largest momentum occurs if there is a parallel process associated with a large logistics-related business establishment in the area, such as a large distribution centre.

With high profitability, stakeholders have clear incentives to start the operational phase as soon as possible to maximize profits and not pass up the opportunity. Various forms of public support can improve/facilitate business profitability. The need for this support is greatest when there is a collective of small-scale users that must be induced to act and collaborate in order to provide a critical mass for operations. There is always a risk that a deadlock may occur in which players are waiting for someone else to invest and act, the advantages of being an "early-mover" are small with few given necessary investments and risks. In such cases, different forms of subsidies may facilitate the development process. The most common type of subsidy is related to the terminal lease or fee. Usually, a variable contract is issued where the terminal operator pays a fee per lift of a container or load unit. This setup must be regarded as a form of subsidy since a newly developed terminal has no or little volume of load units at the start. From an infrastructure owner's perspective, especially a municipality, this is managed by defining a very optimistic forecast of future volumes to be handled at the terminal. From a market perspective, this setup more or less eliminates all form of risk associated with the investments in the terminal for the terminal operator, which means that it has to be regarded as a subsidy. One way of avoiding subsidizing the terminal operator, if desired by the infrastructure owner, would be to have fixed lower and upper fees combined with differentiated handling-based fees, however, such options are rarely exercised.

All terminal developments in Sweden have been based on market potential estimations and analyses, and these are the platform for discussion and negotiation with the Swedish Transport Administration. In the end, the market potential determines whether or not the investments are judged to be socioeconomically positive and thus eligible for possible co-financing from the Transport Administration of up to about 30 per cent of the total cost. There is an analysis made by the Transport Administration of what extent the market potential is "reasonable" but the analysis as such has to be financed by the initiator and investor. However, since many of the terminals are developed in areas where there is no existing intermodal rail-based traffic, it is difficult to identify "reasonable and realistic" medium- and long-term potentials. A more recent problem relates to the situation where new terminals are being developed within each other's catchments areas, however, each terminal might take volumes from other terminals' catchment areas into account in their estimations as it is very difficult to determine where the competitive interfaces are. This issue is especially complicated as there might be several local development processes in a region.

When discussing the problem of excessive densification of intermodal terminals and dry ports it is important to recognize how diverse the markets for terminals are. First and foremost we have a situation where competition, ownership, operation structures, openness, and transparency generate a market demand for more terminals, even if cargo can be handled with existing terminal capacity. This is an important factor that all stakeholders must be aware of when assessing the conditions in which terminals are operating. Another aspect is that there is a general lack of differentiation between terminals. Today, many terminals focus on handling containers. There is a great opportunity to differentiate themselves from other carriers, reefers, biofuels, general cargo, port destinations, international transport, customs clearance, etc. (cf. Bergqvist 2009).

Discussions of terminal density can be expanded to also include various supportive alternative terminal operations where there are strong synergies e.g., efficient use of railway infrastructure, space, handling equipment, and personnel. Today, we can see that several regions with intermodal terminals have attracted the establishment of other types of terminal segments, such as wood chips, logs, sawn wood, and biomass. The result of such an establishment is an overall increased efficiency to all terminals in that area. One reason for this trend is that a competitive location of a rail terminal can be applied to multiple segments and that there is a clear synergy between the terminal operations in the different segments.

2.4.2   Local Enthusiast: Patience and Long-term Commitment

The presence of a local entrepreneur or enthusiast is an important and necessary factor for a successful development of a terminal since the process is often long-lasting and turbulent. In a Swedish context, this enthusiast is usually found at a business development function within the public sector (Bergqvist, Woxenius et al. 2010). Based on all the contingencies and situations that may arise in a development process, continuity and sustainability of key actors are incredibly important and fundamental.

There are numerous aspects that can contribute to friction and problems in the development process and without committed individuals the probability of a successful implementation and operation is limited. A public enthusiast is important in order to maintain a robust development process by good communication with politicians and local decision-makers. Since the development process of intermodal terminals usually lasts over several years, political elections may lead to periods where political decision-making is difficult and sometimes impossible. The presence of enthusiasts is hence an extremely important resource for the development of intermodal terminals. This is particularly evident for small intermodal terminals in the absence of a distinct private investor and project manager that is common for major infrastructure projects.

One way of decreasing the reliance on individuals and local enthusiasts is interaction and collaboration with other individuals and actors. In many cases, there is a strong local expertise in the private and public sector related to the needs and challenges of the regional logistics system. To establish a close and continuous cooperation between the private and public sectors, the local/regional university may play an important role in adding a neutral platform for collaboration where analyses can be initiated and discussed in an academic framework. Collaboration with academics may also lead to a stronger long-term perspective and focus on knowledge creation and exchange.

2.4.3   Financiers

Financing and funding are other major factors in the development process. Besides the level of funding, the number of funders can have a big influence on the collaborative environment. If the involved financiers have different time horizons and rationality in their actions, it risks slowing down the development process. In several Swedish projects, it has been noticeable that public actors such as the Swedish Transport Administration and municipalities have different views on the commercial conditions for terminal operations and how fast investments should be paid off. The experience and the familiarity with working with large infrastructure projects with very long duration and strong regional ties is an important and desirable feature of the financiers so they have an understanding of this complexity.

2.4.4   Localization

A conflict between municipalities on the location of a terminal is a common factor that significantly delays and hinders the establishment and development process. The competition between municipalities is connected to the fact that terminals generally have a larger catchment area than its municipal boundary. An establishment by one municipality generally implies that a future establishment might be very difficult for its neighbouring municipalities in the foreseeable future. At first sight, such a conflict may be perceived as irrational, however, given the long-term implications; it is natural for neighbouring municipalities to be cautious since it may affect the business attractiveness of that region and its different industry clusters. Looking at the conflict from a public decision-maker's perspective, and the fact that the individuals holding political offices have the responsibility to ensure and develop the efficiency of a municipality's transportation systems, it is easy to understand the behaviour. The municipality and the larger region's best interests should coincide in the long term, but there are incentives in the short term that may cause individuals to act and behave differently. The most noticeable "disturbance" is the presence of an election. The feeling of "giving away" job opportunities in a potential terminal establishment to a nearby municipality can be seen as a negative and weak characteristic to voters. The longer the time period to the next election, the harder it is to pursue this strategy successfully. This aspect is important to recognize when planning for key decision points so that the timing does not expose the process to stress that could be avoided or reduced.

2.4.5 The Swedish Transport Administration

This section attempts to illustrate the views and perspectives related to intermodal terminals of the Swedish Transport Administration. The Transport Administration emphasizes that regional inland terminals are important to facilities' future growth of rail freight (cf. Banverket 2010). The Transport Administration has also recognized the importance of local presence and good cooperation between public and private actors to ensure a sustainable and efficient transport system.

One of the Transport Administration's key focuses is the availability of intermodal terminals, which means that they should be open and accessible for all or at least many actors in the transportation system. This means:

> The concept of availability involves the stimulation of broad geographical coverage, transparency, simple predictable decision-making and competitive neutrality in relation to railway operators. Regional initiatives are important for development (translated from Banverket 2010, p. 9).

Some important prerequisites for the establishment of regional intermodal terminals mentioned in research and by the Transport Administration are (Banverket 2010; Bergqvist, Woxenius et al. 2010):

- Location at large production and consumption areas
- Locations which form natural start and end points of goods flows that are linked to major international transport routes
- Should be strategically located in relation to goods flows (intersections of main goods flows, etc.)
- Should be located where it is easy to switch between transport modes and redistribute flows
- Terminal geographic location should allow for efficient train production and attractive lead-times
- The terminals shall be open to the market
- The terminal operator's activities should be transparent to the infrastructure owners and preferably separated from other activities

The Transport Administration has developed strategies related to the development of regional intermodal terminals (Banverket 2010):

- To increase diversity and thus sharpen transportation concepts and cost-effectiveness, all Transport Administration's facilities and other business establishments in which the Transport Administration are co-financiers have the restriction to always be open to all railway operators on competitively neutral terms. This should, if necessary, be agreed upon in special cooperation agreements.

Sub-strategy:

- All new terminals will be designed according to "The functional unit terminal" and the Transport Administration will establish a plan for redesign of older terminals, under this definition.
- Regional initiatives should be identified.

As part of this strategy, the Transport Administration made some observations related to the effective transfer points and logistics parks development:

- By the urbanization taking place, the need for transshipment points near metropolitan areas increases. Logistics parks represent a regional labour political factor. The Transport Administration will be active throughout the planning process around such installations and secure availability of land.
- Rail is efficient for transportation of large volumes over medium and long distances. This applies to costs, environmental impact, and safety. The Transport Administration should concentrate their efforts on technical development, investments, and maintenance related to the major goods flows and to those nodes that are competitive in the long term.

Principle layout of a functional rail terminal, developed by the Transport Administration:

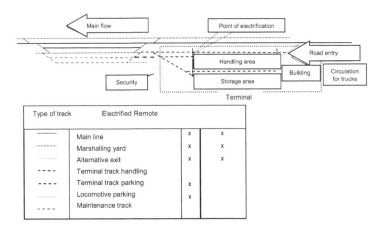

Figure 2.4   The functional unit terminal
Source: Banverket 2010

With the functional unit "Terminal", the Transport Administration refers to:

- Terminal
  - Handling surfaces
  - Load equipment cranes, forklifts, etc.
  - Connecting roads
  - Loading tracks, tracks for temporary parking
  - Buildings
- Transfer Yard
  - Remote controlled with electrified tracks

The transfer yard should have the following characteristics according to the Transport Administration:

- Yard where groups of wagons can be collected and returned with the effort of only the engine crew
- Remote controlled from the central train control
- Electrified if the connecting line is electrified

As far as tendering, ownership, and organization, the Transport Administration has no ambitions to operate cargo terminals themselves (Banverket 2010).

Investments in intermodal terminals are based on a three-level categorization of intermodal nodes in which the Transport Administration may invest more in the higher level than the other two. Regional intermodal terminals with limited flows belong to the lowest level. However, it is possible for terminals to switch between categories / levels if their circumstances change (Banverket 2010).

The Transport Administration has the following restrictions on ownership and contracting:

- The Transport Administration intends to sign contracts with the infrastructure managers and property owners for access rights and control over terminals
- The Transport Administration requires terminal performance based on their sector responsibility:
  - Openness
  - Competitive neutrality
  - Availability

The reason for these requirements or preferences is that they help to open up the transport system to more actors and strengthen the competitiveness of rail transport and rail-based intermodal transport.

2.4.6   Efficient Rail Production

It's easy to forget the importance of efficient rail production when discussing terminal establishment as many of the actors involved in the early stages of the development process often have limited technical experience and knowledge of rail transport and terminal design. Listed below are a number of important aspects that should be considered and analyzed in order to access the efficiency of the achieved rail production related to the terminal operations and associated logistics park:

- Location in relation to superior infrastructure. It is very important to know the limitations, conditions, and opportunities of the infrastructure to which the terminal is connected
- Marshalling. It is important that the terminal and nearby rail infrastructure enables efficient switching/marshalling and prevents unnecessary movements
- Slopes in the area and connecting tracks. This aspect affects the capacity, productivity, and investment needs
- Management of wastewater in the area
- Electrification of the tracks and terminal
- Signalling systems connected to the terminal and the need for switches, etc.

2.4.7   Efficient Terminal Operations

Like the efficiency of rail operations, there are a number of important aspects related to the design and layout of the terminal itself, which strongly affect the efficiency of the terminal operations:

- Paving. The most common surface is asphalt. The disadvantages of asphalt are the more expensive maintenance costs and shorter technical lifetime than concrete stones. Another aspect is that it contributes to increased wear and tear of truck tires, which increases the cost of the terminal operation
- In order not to limit the terminal's capacity too much, it is important to have separated entry and exit lanes and plenty of space in relation to the movements and circulation of trucks
- Another important aspect is the outlet for refrigerated containers/trailers that might benefit from coordinated planning and placement of the terminal's lighting and lighting poles
- It is a clear advantage for the logistics park if all the streets in the park and the connection to the terminal are classified as internal streets. This enables more efficient road haulage as longer vehicles can operate on the roads
- A well-functioning security perimeter around the terminal will improve security and prevent damages, theft, and vandalism. The planning of neighbouring fences and buildings con contribute to this protection while the need for investment in perimeter security is reduced

2.4.8   Independence, Transparency and Openness

Similar to the observations by the Transport Administration, clear ownership and dependencies related to terminals are important for long-term credibility and a functioning transport system. It can be problematic if the terminal operators have direct and specific interests in certain transport flows, as this can affect how the market views the terminal's transparency and the way the operator conducts its business. Although equal treatment with respect to quality and price is guaranteed, there may be commercial and informational barriers that limit competition.

Municipalities often finance terminals, which in itself is not problematic; however, it is an issue when municipalities are involved in the direct terminal operations through part-ownership in the terminal operating company. It is especially problematic if the terminal operating partnership does not work under normal profitability requirements. An additional challenge lies in transparency and the relationship between the municipalities as infrastructure owners if they are also part owners of the terminal operating company.

For the terminal operations, tendering is preferred as it allows for transparency through its public process and the specific conditions. Another advantage is that the terminal owner can continuously monitor these conditions and deviations, and as the ultimate measure, cancel the contract or choose other remedies. These opportunities are very difficult to realize if the terminal operator has "possessory rights" to the terminal in a lease agreement or similar. One extreme of this should be that the municipality is "forced" to buy out its fellow shareholders or tenant in order to achieve the change the municipality wants to implement.

The Transport Administration is another important actor in this context that often finances up to one third of the infrastructure needed related to the rail connection of the regional intermodal terminal. As described earlier, the Transport Administration sets standards of independence and transparency, but the wording is quite vague and unclear. From the perspective of the Transport Administration, there is a very limited degree of monitoring and reporting related to the demands put forward in the cooperative agreement once operations start. An overall risk of terminals that are not considered independent and transparent is the "market" believes that there is a need for more terminals, which can lead to over-establishment of terminals and diseconomies of scale.

2.4.9   Tendering and Agreements

Today, many infrastructure owners, such as the municipalities in Umeå and Falköping and the state-owned company Jernhusen choose to "procure" terminal operations by tendering. The interest from the market has been great. The tendering procedures and tendering documents have required better frameworks to be developed related to risk, service, contract periods, contract options, leases, marketing of the area, etc. Traditionally, the operation of many terminals was given to the actor who first reported interest. Perhaps the most important element of these tendering processes is that the infrastructure owner is presented with new ideas as the bidders present their quotes and concepts. The aspect of tendering has increased the level and pace of innovation within the segment of terminal operations and logistics park activities and services.

2.5   Current Trends and Challenges

Despite its historical development and good geographical coverage of the hinterland transport system connected to the Port of Gothenburg, the system still has possibilities to develop new segments. The segment of semi-trailers is a possibility that could substantially increase the volumes in the Scandinavian rail shuttle system. The semi-trailer segment poses some significant challenges since it is very different from containers in many aspects (Woxenius and Bergqvist 2011). A market-share of about 20 per cent for the semi-trailer segment could increase the volumes of the rail shuttle system by about 100 per cent (Bergqvist 2009).

Figure 2.5   Intermodal transport of semi-trailers
Source: Photo © Fredrik Bärthel

Currently, almost all dry ports and rail shuttles in the Scandinavian rail shuttle system have the same single idea of transporting containers to and from the Port of Gothenburg. There are great possibilities for differentiation and specialization related to the segments of semi-trailers, swap-bodies, bulk, stripping and stuffing, reefers, express goods, specific industries (e.g., furniture, groceries), etc. Other noteworthy possibilities now starting to emerge include the interest by other sectors such as biofuel, round timber, etc.

2.6   Conclusions

Terminals in Sweden are traditionally a local and regional matter, which generates local attention and interest. They contribute to local support by decision-makers and acceptance that can facilitate modal shift from road to rail. However, there are also a number of important factors related to this observation. A very common example is the emergence of conflict at the regional and local level, e.g., several neighbouring municipalities want to establish an intermodal terminal in a market that normally supports one. When such conflicts appear, there is a need for an actor/authority that can take greater responsibility for the overall logistical efficiency. A weak commitment and lack of initiative in these matters could delay the development of the hinterland transport system.

The factors and issues addressed in this chapter related to the terminal development process are illustrated in Figure 2.6. All factors are of course interconnected; however, the links in the figure illustrate the strongest relationships. From an evolutionary perspective, these linkages are especially interesting, as the strength of these relationships constitutes the development process and ultimately the efficiency of the terminal and related intermodal transportation.

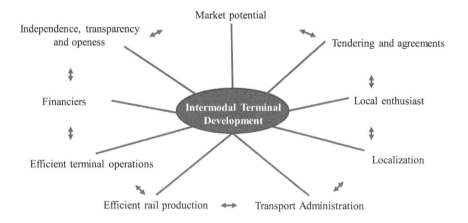

Figure 2.6   Important factors related to intermodal terminal development

Regional intermodal terminals and dry ports have an important role in ensuring a competitive and sustainable Scandinavian transport system in the future. The most important prerequisites for the future development of intermodal transport are capacity at the endpoints and the intermediate infrastructure. Herein lies both the investment and expansion of new infrastructure and changed priorities within the existing infrastructure. Related to the development of regional intermodal terminals and dry ports and sometimes adjacent logistics parks, is the importance to recognize the need for a continuous process of creativity, innovation, and competence.

2.7   Acknowledgements

The author would like to thank all the interviewees who contributed with their time and provided valuable input to the research. The research was conducted with financial support from the Interreg IVb North Sea Region-funded Dryport project (cf. www.dryport.org).

2.8   References

Banverket. 2010. Inriktning för godstransporternas utveckling. v. BVStrat 1003, Samhälle och planering.

Bergqvist, R. 2007. *Studies in Regional Logistics: The Context of Public–Private Collaboration and Road–Rail Intermodality*. Logistics and Transport Economics, Department of Business Administration. Göteborg: BAS.

Bergqvist, R. 2008a. Organisatoriska processer vid etablering av kund- och agentinitierade intermodala transportsystem, in *Nya aspekter på intermodala transportkedjor: Tre förstudier*, edited by A. Jensen, Göteborg, SIR-C Rapport.

Bergqvist, R. 2008b. Realizing Logistics Opportunities in a Public–Private Collaborative Setting: The Story of Skaraborg, *Transport Reviews,* 28(2), 219–237.

Bergqvist, R. 2009. *Hamnpendlarnas betydelse för det Skandinaviska logistiksystemet*, Handelshögskolan vid Göteborgs universitet. Göteborg: BAS

Bergqvist, R., Falkemark, G. and Woxenius, J. 2008. *Establishing Intermodal Terminals*. Nectar Logistics and Freight cluster meeting, Delft 27/28 March 2008.

Bergqvist, R., Falkemark, G. and Woxenius, J. 2007. *Etablering av kombiterminaler*, Meddelande 124. Göteborg: Department of Logistics and Transportation, Chalmers University of Technology.

Bergqvist, R., Woxenius, J. and Falkemark, G. 2010. Establishing Intermodal Terminals, *World Review of Intermodal Transportation Research,* 3(3), 285–302.

Port of Gothenburg 2011a. *Clear Environmental Gains from the Port of Gothenburg Rail Investment*. Available at: http://www.portgot.se/prod/hamnen/ghab/ dalis2b.nsf/vyPublicerade/93D1590B47738CD7C1257505003702D1?OpenD ocument [Accessed 2011.02.08].

Port of Gothenburg 2011b. *Rail Services May 2011*. Gothenburg: Port of Göteborg AB.

Roso, V. 2006. Emergence *and Significance of Dry Ports*. Division of Logistics and Transportation. Göteborg, Sweden, Chalmers University of Technology: 43.

Trafikanalys 2010. Bantrafik 2009. Trafikverket

Woxenius, J. and Bergqvist, R. 2011. Hinterland Transport by Rail: Comparing the Scandinavian Conditions for Maritime Containers and Semi-trailers, *Journal of Transport Geography,* forthcoming.

Chapter 3

Dry Ports:
A Concept or a Reality for Southeast Drenthe?

Johan Gille and Jeroen Bozuwa

3.1   Introduction

The concept of a dry port has been receiving substantial attention over the past years, both from researchers/academics, policy makers and logistics operators (Roso and Lumsden 2010, Rodrigue et al. 2010, Harmsen 2010). Especially, the latter have been seeking to improve the efficiency of transport chains and making most efficient use of the available space in seaports and in hinterland sites. However, it appears that the term dry port is sometimes misused, becoming a marketing instrument rather than a clearly defined concept that is implemented.

The purpose of this chapter is to outline the various definitions and functionalities of dry ports, and to show the consequences if these are applied to the case of Southeast (SE) Drenthe. Firstly, an overview of definitions and visions made in various studies is presented. Secondly, relevant services at a dry port are described, and the expected contribution to economic development and to regional transport policy objectives of such facility is presented. Finally, using Southeast Drenthe as a case study, the consequences of implementing a dry port are discussed in a regional economic context.

This work builds on a study conducted on behalf of the municipalities of Emmen and Coevorden and the Province of Drenthe as preparation for their work under the Interreg IVB project Dryport (Ecorys 2009).

3.2   Definition of a Dry Port

This section will look into definitions used, types of existing dry ports and the relevance of distance.

Over the past years, several articles have been published addressing the dry port concept (Roso and Lumsden 2010, Rodrigue et al 2010, Van den Bossche and Gujar 2010, Frost 2010). Often the definition is subject to discussion and various outcomes have been found. For example Roso and Lumsden (2010) and Frost (2010) specify a dry port as an inland intermodal terminal that is connected directly to seaport(s) with high capacity transport means, frequently rail, so that customers can drop off/pick up their units as if directly to a seaport.

Roso and Lumsden (2010) evaluate several dry ports around the world but limit their study to those having railway connections. Another study by Rodrigue et al. (2010) uses the term of inland port rather than dry port, as they consider this term more appropriate, although they admit that variations are also based on a certain level of semantics. For example the term "dry" might suggest that inland terminals serviced by barges are excluded. Furthermore the question raised is whether a dry port or inland port is an inland freight hub or merely a regional distribution centre. In their view, three factors are of importance: containerisation, a dedicated link between seaport and dry port, and massification. FDT (2007) use a broader definition for a dry port, including also the services that are typically provided in the seaport (e.g. customs clearance, container maintenance and repair, empty depot):

> A dry port is a port situated in the hinterland servicing an industrial/commercial region connected with one or several ports by rail-, road- or inland water transport and is offering specialised services between the dry port and the overseas destinations. Normally the dry port is container and multimodal oriented and has all logistic services and facilities needed for shipping and forwarding agents in a port.

For our case study, the following factors are relevant:

1. The existence of a clear link in terms of freight flows with one or more seaports,
2. A basic level of integration of logistics, customs and IT services;
3. Multimodal access;
4. Provision of services to the industrial/commercial region in which the dry port located,
5. Optional, a function as an inland hub for long distance multimodal transport.

Before discussing the objectives of a dry port, some basic characteristics of different types of dry ports are explained (see Figure 3.1):

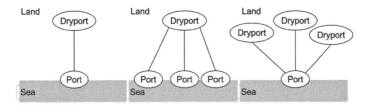

Figure 3.1   Different types of dry ports
Source: FDT 2007, NTU (www.ntu.eu)

- A single dry port servicing one seaport;
- A single dry port servicing multiple seaports;
- Multiple dry ports servicing the same seaport.

Dry ports can be distant, at mid-range or at short-range to seaports (FDT 2007). As mid-range and short-range dry ports are close to the seaport and transport distances are relatively small, inbound and outbound flows are mostly handled by road transport. Distant dry ports are located further in the hinterland, so the transport distance between the seaport and the dry port is much larger. Inland shipping and rail become more competitive on these longer transport distances. One could argue whether the short-range and mid-range categories are to be called dry ports when the main transport mode between the seaport and the dry port is provided by road transport rather than higher capacity transport modes like rail or barge, as some definitions require (Roso and Lumsden 2010). However, the definition chosen here does not differentiate distance, thus leaving this possibility open. In the context of the Dutch geography and transportation network, we estimate that short-distance dry ports are located within about 50 kms from the seaport, mid-range dry ports at 50–150 km and distant dry ports located beyond this distance. In the Netherlands, the planned container facility at Alblasserdam may qualify as a short range dry port – and Rodrigue et al. (2010) would consider this a satellite, although they define Venlo to be a satellite at a distance of 160 km from Rotterdam – while Southeast Drenthe would fall into the category of distant dry ports category. The same applies to for instance Kisarawe, serving as a dry port to Dar Es Salaam alleviating the seaport from road congestion (Ecorys 2011).

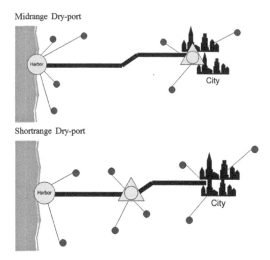

Figure 3.2   Mid-range and short-range dry port
Source: Dryport 2009

FDT (2007) has investigated advantages and disadvantages of dry ports in the different distance categories (see table below). The study concludes that all three types of dry ports increase inland accessibility, strengthen multi-modal solutions, avoid traffic bottlenecks and reduce pollution.

Table 3.1   Dry port advantages and disadvantages

	Close dry port	Middle range dry port	Distant dry port
Conditions	• Transit activity dominant in the seaport • There is a need due to the lack of space at the seaport	• High volume customers • Rail link between seaport and market	• Rail link between seaport and market
Location level	• Decongestion of the city access • Reduction of pollution • Increased intermodal transportation	• Region attracts industries • Reduction of pollution • Increased intermodal transportation	• Acquiring new hinterland of the seaport in consideration • Reduction of pollution • Increased intermodal transportation
Infrastructure level	• Reduction of road maintenance cost • Increase of rail maintenance cost • Reduction of cost for road infrastructure development • Increase of cost for rail infrastructure development	• Reduction of road maintenance cost (in case of pay roads, reduction of profits) • Increase of rail maintenance cost	• Reduction of road maintenance cost (in case of pay roads, reduction of profits) • Increase of rail maintenance cost
Transport level	• Light reduction activity for road carriers from/to seaports • Reduction of congestion and waiting time for transport operators • Increase of transit time • Increase of handlings	• Reduction activity for road carriers from/to seaports • Reduction of congestion and waiting time for units • Decrease of transport costs • Coordination with rail passenger traffic	• Light reduction activity for road carriers from/to seaports • Reduction of congestion and waiting time for units • Decrease of transport costs • Coordination with rail passenger traffic
Logistical level	• Increased inland access and city distribution • Invitation for the use of intermodal solutions	• Increased inland access • Decrease of costs	• Increased inland access • Possibility to choose between ports • Decrease of costs
Customers viewpoint	• Raise of costs at the beginning • Decrease of costs in the long run • Reception of units closer to their own geographical location	• Easy access to seaport • Decrease of costs • Slight increase of transit time	• Easy access to seaport • Decrease of costs • Increase of transit time (or decrease depending on the country of interest, on its road infrastructure quality level, and on distance to cover

Source: FDT 2007

3.3 Services Provided in a Dry Port

Discussions among policy makers on dry ports mainly deal with the steps to be taken to transform basic inland transhipment points into areas where more services are offered (SNN 2004, Horst 2008, Ecorys 2009). Conventional inland terminals provide the transhipment of goods between modes as a basic service. Following the definition chosen here, full-service dry ports include many other functions such as storage, consolidation, depot storage of empty containers, container maintenance and repair and custom clearance. This list is neither exhaustive nor does each dry port require all of these services to be available, as long as "all logistic services and facilities needed for shipping and forwarding agents" are in place (Ecorys 2009).

Another factor of relevance for becoming a dry port is the so-called "extended gate" principle (Visser et al. 2009). This implies that the "gates" of the seaport are extended into the inland and that the shipper or forwarder sees the dry port as an adequate interface towards the seaport and the maritime shipping lines. This not only means that similar services are in place as are offered in seaports, but also that there is a seamless integration of these services and especially of information exchange between the dry port and the seaport(s) to which it connects.

By creating an "extended gate" at inland terminals, containers can be transhipped directly by rail or barge from the seaport terminals to and from the hinterland. In the "extended gate" concept, containers are moved directly from the sea-going vessel to and from the inland terminal under the responsibility and customs license of the seaport terminal operator. After clearing customs at the inland terminal the containers are transported by truck, barge or train to their final destination. Export boxes can also be delivered at the inland terminal and are, at that moment, virtually loaded on board of the sea-going vessel, thanks to the proper ICT-connections. This co-operation enables the inland terminal to utilise the capacity at the marine terminal in the most efficient way and diminishes the pressure on motorway capacity. The inland terminals are in constant contact with the terminals in the seaport. The issuing and processing of orders, invoicing and customer reporting all takes place directly from the computer system of the seaport operator.

According to the Interreg IVB North Sea Dryport project, a dry port should provide the following layers of services (Dryport 2009):

- Intermodal transport and handling of goods;
- Information handling;
- Load unit handling;
- Customs;
- Logistics services.

These five layers can be seen in order of development level, the lowest being the transhipment of goods itself, with extra complexity added for each subsequent layer (i.e. handling of information and so on) (see Figure 3.3).

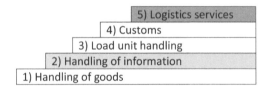

Figure 3.3   Dry port functionalities
Source: Interreg IVB North Sea project Dryport.

FDT (2007) developed a structured questionnaire in order to evaluate the importance of the dry port functions. The responses from the 69 participants revealed that from the respondents perspective dry ports should offer the same functions as a seaport, in order to have a competitive advantage (Figure 3.4).

From the point of view of seaport container terminal managers, a returning issue is the repositioning of empty containers (Frost 2010). An important way to improve the efficiency of container transport using dry ports is to cooperate with seaports in a system that uses the dry port capacity for repositioning empty containers. In the Netherlands it is considered that barge operators and terminals, if given the right information regarding regional transport patterns and having their own network covering (part of) the country are able to get empty containers to the right place at relatively low cost (Ecorys 2009). Shipping companies often do not have a good insight in the merchant haulage equipment flows of their containers due to the fact that in merchant haulage, the shipper uses a forwarder to organise the transport – compared to the carrier haulage streams where the container shipping

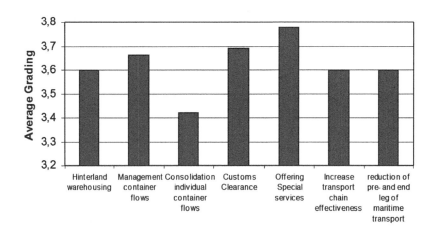

Figure 3.4   Necessary functions of the dry port (scale: 1 = not important, 5 = very important)
Source: FDT 2007, NTU (www.ntu.eu)

company organises the landside transport himself – and the forwarder in turn uses a transport company for the transport. As a result the container shipping companies do not always know the whereabouts of "their" containers. In the Netherlands/ Western Europe the Flemish Institute for Logistics (VIL 2005) assessed that the greatest part of the hinterland transport comprises of merchant haulage streams. In the case of Southeast Drenthe this would mean that a dry port development could improve the efficiency of container transport to/from the main seaports if they would participate in a system for the repositioning of empty containers between inland terminals (SIKZNEB 2005). Based on current haulage flows, these inland terminals would comprise rail and inland water terminals in both The Netherlands and in Niedersachsen, Germany.

3.4   Added Value of a Dry Port to the Local Economy

FDT (2007) discusses how the potential development of an inland intermodal facility into a dry port can lead to an increased economic development in the adjacent area. By expanding available services at an inland multimodal terminal, related services could develop from the dry port and thus create added value in the region; thus giving positive economic impact in the region where the dry port is located.

Figure 3.5 shows an example of some different levels of service functions that can develop when establishing a dry port (the two inner circles indicate the minimum level of services which should be available at a dry port according to the UN report, 2006). The figure shows the potential expansion of service functions as a result of the evolution of an inland multimodal terminal into a dry port or of the development of a dry port from scratch.

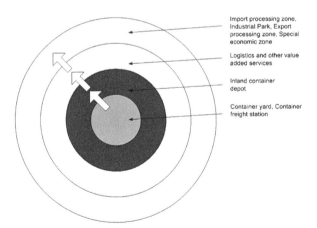

Import processing zone, Industrial Park, Export processing zone, Special economic zone

Logistics and other value added services

Inland container depot

Container yard, Container freight station

Figure 3.5   Potential expansion of functions towards a dry port
Source: UN 2006

Examples of this development can be found in Venlo, Germany, where a significant number of traders is located adjacent to a large cluster of the greenhouse sector, contributing to additional freight flows through the existing dry port. Here as elsewhere, it may be a question of "what must come first". A dry port can be a trigger to attract other economic activities, yet a minimum volume of cargo flows must already be available to make the principle investment in a dry port sound and feasible. In summary, dry ports can:

- act as freight storage and distribution centres for the international market (if customs services are in place – inland customs clearance posts)
- be used to cut import costs and improve the management delivery times and production processes.
- help to reduce the pressure saturated and high costs locations (e.g. for storage of empty containers).

3.5   Contribution to Achieving (EU) Transport Policy Objectives

All dry port definitions have in common that they are based on a seaport being directly connected with an inland intermodal terminal (the dry port), where goods in intermodal loading units can be turned in as if directly to the seaport. Between the seaport and the dry ports relatively large amounts of goods in intermodal loading units can be shipped as if the terminal was positioned directly at the seaport. From the point of view of modal shift policies, dry ports offer a potential by aligning large freight volumes on a single corridor between the seaport and the dry port. If the volume is sufficient, a shift from road to more energy efficient and less environmental harmful transport modes becomes feasible. .In addition a dry port can relieve adjacent seaport roads as well as seaport cities from congestion caused by port related road transport flows, making goods handling more efficient and facilitating improved logistics solutions for shippers in the seaport's hinterland. According to Roso and Lumsden (2010) this is the least recognised advantage of a dry port, but from the perspective of the seaports it may be a crucial condition when considering their involvement in dry port development. On the other hand a dry port mainly servicing a seaport to solve its congestion problem may be vulnerable to market fluctuations. Rodrigue et al. (2010) mention that in the case of declining traffic or capacity expansion at the seaport, the seaport terminal may recapture the activities performed at the satellite dry port.

From interviews held with stakeholders in Southeast Drenthe and other parts of the Netherlands we conclude that strengthening multi-modal solutions and avoiding traffic bottlenecks are the most important dry port advantages (Ecorys 2009). The urgency is furthermore emphasised by the fact that the share of road transport has increased over the past years and, if this trend continues, will cause increased congestion problems in the future.

3.6   Case Study: Southeast Drenthe

Can the dry port concept, as presented in the previous section, be realised in the region of Southeast Drenthe, in the Netherlands? This region, with the two main economic centres: Emmen and Coevorden, is located in the northeast of the Netherlands, along the transport corridor between the seaports of Amsterdam and Rotterdam, and Northern Germany/Scandinavia (see Figure 3.6). The latter already has a large logistics park available including a multimodal container terminal. Both Emmen's and Coevorden's industrial areas are accessible by highway and rail. Additionally, Coevorden has access to inland the inland waterway network, albeit with a limited capacity (CEMT-II, allowing vessels up to 600 tonnes). A pilot using vessels of 36 TEU on the canal was carried out in 2009 (Schuttevaer 2009). The Coevorden logistics park comprises international rail access to the German railway network via the Bentheimer Eisenbahn.

In order to assess the potential of SE Drenthe to evolve into a dry port, the case study addresses the following questions:

- Are the volumes of freight being transported between main seaports and the region, or passing along the region, sufficient to allow for multimodal freight services concentrated in a dry port?
- Are the required infrastructure and services in place?
- How do stakeholders envisage the potential of a dry port in this region?

The answers to these questions provide conclusions on what actions are to be taken regionally for the dry port to be able to evolve.

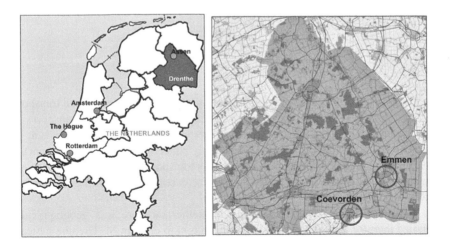

Figure 3.6   Geographic location of the Southeast Drenthe region
Source: Ecorys 2009

3.6.1   Freight Flows

First we have assessed the freight volumes potentially relevant for a dry port in SE Drenthe from various perspectives, notably:

- Freight flows from major Dutch seaports to the region and vice versa, as well as freight flows between the region and major German North Sea seaports.
- Freight flows from major seaports to international hinterland destinations beyond SE Drenthe which may be routed via a hub in this region, as well as vice versa flow from German North Sea seaports to the Northern Dutch hinterland beyond SE Drenthe.

3.6.1.1 Freight flows to/from Emmen/Coevorden

The following table shows the captive volumes of freight by commodity group and transport mode that are shipped to and from the region.

Table 3.2   Freight flows to/from SE Drenthe by commodity (*1000t)

	Road	Rail	IWT
Chemicals	2.214	160	11
Containers	324	187	0
Dry bulk	6.651	0	62
Liquid bulk	330	0	1
General cargo	4.455	5	3
Total	**13.974**	**352**	**77**

Source: Ecorys 2009

　　Freight flows to/from the region are dominated by road. Within road transport, the volumes are primarily dry bulk (48 per cent of all captive road transport). In rail, the container segment is largest at 53 per cent, followed by chemicals at 45 per cent. Inland Water Transport (IWT) mainly concerns dry bulk (81 per cent).

　　However, only about one million tonnes is transported between the region and Dutch seaports, while the majority has origins/destinations elsewhere within the Netherlands.

　　Statistical data on rail freight in the Netherlands are less geographically detailed than for road. Current data on rail freight to or from SE Drenthe indicate that practically all rail freight (see Table 3.2) is shipped to/from the Rotterdam Rijnmond area, which is logical given the currently operated rail services to/from the trimodal Euroterminal Coevorden.

The volume of barge transport to/from the SE Drenthe region itself is very small and originates from various parts of the country as well as from Germany, though it is hardly related to origins in seaports.

Freight flows to the region originating in Germany and vice versa have also been analysed. Regarding destinations within Drenthe, German statistics do not give a detailed breakdown, but only provide total volumes, amounting to about 2.2mt (2007 data). Less than one per cent of this originates from German seaports.

Overall it can be concluded that some 16mt are transported to/from the region by the three modes under study. However, only a minor part of these flows is originating in or destined for seaports. Nevertheless a logistics service centre (dry port) may be able to capture part of these flows, thus not limiting its activities to seaport connected cargoes only.

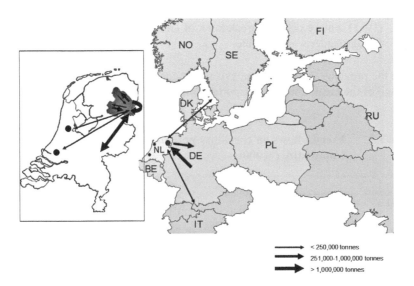

Figure 3.7   Freight flows to/from SE Drenthe by commodity (*1000t)[1]
Source: Ecorys 2009

3.6.1.2   Freight flows on corridors along the region

Other potential freight flows are related to cargo that does not have its origin or destination in the region, but are in transit along corridors that pass the region,

---

1   It is noted that freight flow data is presented in tonnes. In container transport, often the unit TEU is also used (20-foot equivalent Unit). However, as statistical data on freight flow origins and destinations is provided in tonnes, and as we also include non-containerised flows (dry and liquid bulk, general cargo), the analysis is performed in tonnes.

and thus might be handled between modes at a dry port located strategically along these corridors. For a dry port it will be more difficult to intercept part of these flows than it will be to capture flows destined for its own region. On the other hand these flows are potentially relevant if the dry port is to develop as a hub serving long distance corridors. The analysis has considered corridors in two directions:

- Eastbound: from Dutch seaports Rotterdam and Amsterdam to the Northern German, Scandinavian and Eastern European hinterland;
- Westbound: from German North Sea ports to the Northern Netherlands.
- A problem in analysing the origin–destination (OD) data for international destinations was that Dutch statistics only provide foreign destinations at country level. This implies that also destinations e.g. in Southern Germany are included in the analysis, although most likely they will not be served through a Northern Dutch dry port. Only for inland waterway transport (IWT) a more detailed breakdown of German destinations at the level of Bundesländer was available.

As in the previous section the analysis was made for three modes. The following table gives the breakdown of transported volumes by mode and commodity type.

Table 3.3   **Freight flows on corridors potentially passing SE Drenthe by commodity (*1000t)**

	Road	Rail	IWT
Chemicals	1.150	580	326
Containers	4.096	604	0
Dry bulk	739	11.621	1.343
Liquid bulk	60	22	294
General cargo	4.860	1.573	322
Total	**10.905**	**14.400**	**2.285**

Source: Ecorys 2009

The mix of commodities transported on corridors potentially passing SE Drenthe differs from that of captive cargoes (having its origin or destination in the region). For road transport, containers and general cargo dominate, with 38 per cent and 45 per cent respectively. In both rail and IWT, the dominant commodity is dry bulk.

Clearly, the modal split for non-captive flows (e.g. flows on corridors along the region) is more balanced than for the captive flows. The modal split of non-captive flows is in favour of rail (53 per cent) and road (38 per cent), leaving some nine per cent for IWT. For captive flows we can observe that road transport is dominating (see Figure 3.8).

The OD analysis has shown that the volumes decrease as distance from Dutch seaports increases. By way of example Germany generates 96 per cent of the road flows, while the Scandinavian countries cover the remaining four per cent. Volumes to Poland and Russia are negligible. In rail, this is even more pronounced, with Germany being responsible for 99 per cent of the volumes. In difference to road transport, not Scandinavia but Poland is the second most important destination, albeit at a marginal one per cent of total volumes (2008 data). Figure 3.9 shows results for rail transport.

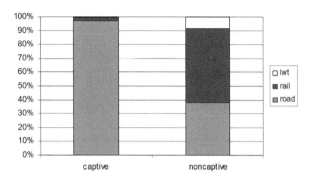

Figure 3.8   Freight modal split captive vs non-captive flows
Source: Ecorys 2009

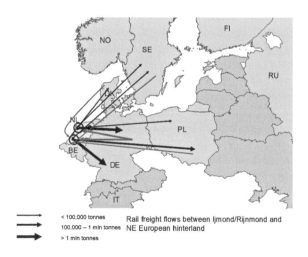

Figure 3.9   Rail freight flows between IJmond/Rijnmond and NE European hinterland in 2006
Source: Ecorys 2009

For inland waterways, Dutch statistics are more detailed and present German destinations by Bundesland. In the analysis we have included the Northern German Bundesländer as well as all Bundesländer in the Eastern part of Germany. Volumes by IWT from Rotterdam/Amsterdam to Northern and Eastern Europe are rather modest with 2.3mt. This is 100 per cent related to Germany, with only very marginal volumes to other northeastern European countries. Within Germany, (Western) Niedersachsen is the most important region, followed by Sachsen-Anhalt. This applies for both directions.

Vice versa, flows from the German seaports to regions in the Netherlands are relatively small (0.2mt from Hamburg, 0.7mt from Bremen), but from land-based sources in Northern Germany, more than 10mt are being transported to the Netherlands, mainly to the central and southern parts. The dry port SE Drenthe is centrally located on this broad corridor, and might be able to serve part of these flows. However, as freight flows are scattered, and freight is deliver mainly by road (88 per cent), it will be rather difficult to attract these.

For the non-captive freight flows, it can be concluded that sea ports are important source regions. Hinterland destinations can be rather scattered, although it is clear that the nearby German Bundesländer are the most important regions. This provides an attractive basis for a dry port in Southeast Drenthe.

Total non-captive volumes transported along the corridors that were analysed amount to some 27mt. It must be noted that this includes road and rail freight to all German destinations from all parts of the Netherlands Therefore the actual potential for a Southeast Drenthe dry port will be significantly lower than this volume.

The Dutch Ministry of Transport assumes that by 2020 freight transport in the northern region of The Netherlands will have increased by 48 per cent (SNN 2004). According to SNN (2004) the strongest increase will take place in road (49 per cent) and inland waterway transport (60 per cent).

The perspective of freight flows shows a good potential for developing a dry port in SE Drenthe. About 1mt of freight is already being transported between seaports and the region itself, with a tenfold volume already transiting the region on its way between seaports and longer distance destinations.

3.6.2   Infrastructure and Services

This section elaborates on the aspects of necessary services and infrastructure development for a dry port.

3.6.2.1   Infrastructure

In order to develop a dry port, good multimodal infrastructure should be in place. This can be evaluated objectively, but also shippers should be asked about their views with regard to their needs. Our analysis identified that the perception among industry is that the waterway access to Coevorden is insufficient with CEMT-II (600t), despite the fact that pilots have shown that feasible container shipping is possible.

With regards to rail, an asset of the region is its direct border crossing connection (the Bentheimer Eisenbahn). At the same time several other rail and IWT hubs in the region may be competing for cargo with Southeast Drenthe (e.g. Veendam, Meppel).

The road network giving access to the region can be considered as good; especially the recent upgrading of the A37 has improved the East–West connection (Randstad–Zwolle–Emmen/Coevorden–Bremen–Hamburg and Scandinavia). Plans upgrading the road transport network within the region are under consideration. Interviews with regional industry stakeholders (shippers, logistic service providers) revealed that they widely appreciate the high accessibility for road transport in the region and low level of congestion.

Coordination between shippers and transport companies is required to ensure consolidation of freight volumes, in order to create sufficient demand to establish a dry port. As there are several other logistics hubs located in the north of the Netherlands (Meppel for IWT, Veendam for rail), some coordination between terminals may also be necessary to avoid segmentation.

3.6.2.2 Services

The availability of know-how on customs procedures is believed to be an important potential asset for the region. It is generally acknowledged by the interviewees that there is a shortage of highly trained and experienced experts in the field of custom procedures.

Furthermore, local stakeholders believe that the presence of logistics expertise in the region (business and research) should be more widely promoted. In this respect the presence of Stenden University is seen as a contributing factor to expand knowledge on logistics. Remarkably, stakeholders perceive that there is no structural cooperation on specific research questions, innovative pilot projects or other multi-annual research programmes.

Finally, shippers consider the price and quality of service as generally more important than the location of the supplier. The most important quality criterion is the reliability of transport services, which is perceived even more important than speed. Interviewed shippers and logistic service providers argued that (the possibility for) offering high frequent services is a good strategy to increase reliability; particularly in combination with using multimodal transport connections.

One interesting comment raised by companies interviewed was that they do not often look across the border beyond the Europark for providing logistics services. The amount of service providers on the German site of the border appears to be limited, which could offer an opportunity for Dutch logistic service providers to serve shippers on the German side of the border via Emmen-Coevorden.

3.6.3   Stakeholder Views

Any development requiring participation of private stakeholders cannot be successful if they are not supporting it. Implementing a dry port clearly is such a development, requiring commitment of service suppliers, shippers as well as all relevant authorities. Interviews with stakeholders from both the Emmen-Coevorden region and from within the ports of Rotterdam and Amsterdam revealed that unanimity in opinion irrespective of the geographic location of stakeholders.

First of all, stakeholders indicated the need for a substantial cargo volume to make multimodal services feasible. In order to develop as a logistic hub or hot spot it is vital that sufficient mass of transport flows, particularly for rail and IWT are generated in the region. This means that according to stakeholders the use of inland waterway transport as an attractive option the inland port of Coevorden should be accessible for ships with a minimum loading capacity of approximately 1,200 tonnes.

Secondly, beyond volume a certain frequency of services, connecting the region to other hubs is required. This is especially relevant for rail connections. For operating frequent rail services (according to stakeholders) a minimum of 2–3 services per month are needed. Joining forces between Emmen and Coevorden would be a strong case to create the required volumes. Interestingly, current frequencies are higher than 2/3 times per month, which indicates that apparently some stakeholders are not aware of the actual market situation. Stakeholders though emphasised the availability of cross-border rail connections.

Thirdly, coordination is required to bundle freight flows. Such coordination demands a very good understanding of the shippers' varying logistic chains and between potential transhipment centres (Emmen, Coevorden, Meppel, and Veendam). The presence of logistic service providers in Emmen-Coevorden, operating from different terminals can help to expand the network and coordinate freight flows in order to generate sufficient volumes.

Fourthly, stakeholders mention close cooperation with seaports such as the concept of "extended gates" used by ECT. ECT operates (or participates in) inland terminals with multimodal facilities (rail, barge) in Venlo (in the southeast of the Netherlands, near the German border), Duisburg-DeCeTe (Germany) and Willebroek (Belgium) and recently started a co-operation with CTVrede - Steinweg in Amsterdam and Moerdijk. Concerns about shareholder influence may not be necessary, as can be seen by the example of RSC which is owned by DB, but acts as an independent company.

Finally, shippers need to be convinced to use the dry port rather than arrange their own transport. When trying to bundle regional freight flows logistics operators require good knowledge of the freight flows of companies (shippers) operating in the region. Companies have very specific demands that can be hard to meet. For larger companies in the region (e.g. Teijin and DSM) combining freight flows with those of other companies might not be attractive as they are of sufficient scale to demand their dedicated services and the combining these with others might only add additional risks to the logistics chain.

3.6.4   Recommendations for SE Drenthe Dry Port Development

On the basis of the freight flows analysis, the review of availability and quality of infrastructure and services as well as opinions of interviewed stakeholders, the following SWOT analysis has been constructed.

Table 3.4   SWOT-analysis SE Drenthe dry port

Strengths	Weaknesses
• Available space • Good rail and road accessibility • Direct border-crossing rail connection (Bentheimer Eisenbahn) • Tri-modal terminal site • Not congested	• Not along main East–West corridors • Lack of high-skilled labour • IWT accessibility is poor • Low capacity on rail-link between port of Rotterdam and Coevorden/Meppel • Rail track length at Europark is limited in length
Opportunities	Threats
• Develop IWT (including terminal equipment) • Bundling of flows (especially SME) • Modal shift to rail/IWT • Cooperation with seaports both in NL and GE • Use of operators that are also active in other regions • Promote as modal shift point in green corridor	• Competition from nearby terminals • Increase of scale, especially regards IWT

Source: Ecorys 2009

Discussions and further evaluation of the SWOT with about 50 regional stakeholders delivered the following six recommendations to further develop the idea d implementing the dry port concept in Emmen-Coevorden:

1. Develop an integral vision and action programme on the desired logistic development/ ambitions;
2. Establish commitment and cooperation of the main stakeholders in the region;
3. Appoint a coordinator to organise and monitor this development and coordinate actions;
4. Promote the dry port among potential shippers, transport operators and all stakeholders;
5. Strengthen the relation with other dry ports and seaports;
6. Develop the required (knowledge and physical) infrastructure.

3.7   Conclusions

Multiple definitions of the dry port concept exist, ranging from mere marketing of inland terminals to full-fledged operations covering all services at the same level as seaports. The differences in understanding are also shown in discussions with stakeholders in the case of SE Drenthe.

For the further development of a dry port in SE Drenthe, the stakeholders in the region will need to clearly define which services and infrastructure they want to integrate into the dry port concept. Coordination between shippers and transport suppliers is needed as well as between the Emmen/Coevorden site and other freight hubs in the region. Also it must be acknowledged that the region is not located along major West–East corridors (Rhine, Betuweroute) and will need to target other flows. Despite these reservations, the analysis has shown that the volumes identified are still substantial.

3.8   References

Bossche, M. van den and Gujar, G. 2010. Competition, Excess Capacity and Pricing of Dry Ports in India: Some Policy Implications. *Journal of Shipping and Transport Logistics*, 2(2), 151–167.

Dryport 2009. *What is Dryport?*. Available at: http://www.dryport.org [accessed 29 March 2011].

Ecorys 2009. Dryport Southeast Drenthe: Strengthening the Logistic Hub. Rotterdam.

Ecorys 2010. Landelijke Markt en Capaciteits Analyse (LMCA) binnenhavens. Rotterdam.

Ecorys 2011. *Pre-feasibility Study, Review of PPP Options and Optimum Option for Establishment of the Kisarawe Freight Station*. Rotterdam, February 2011.

FDT 2007. *Feasibility Study on the Network Operation of Hinterland Hubs (Dryport Concept) to Improve and Modernise Ports' Connections to the Hinterland and to Improve Networking*. Aalborg.

Frost, J.D. 2010. The "Close" Dry Port Concept and the Canadian Context.

Rodrigue, J.P., Debrie, J., Fremont, A. and Gouvernal, E. 2010. Functions and Actors of Inland Ports: European and North American Dynamics. *Journal of Transport Geography*, 18(4), 519–529.

HHLA and POLZUG 2010. *Intermodal Inaugurate High-Performance Hinterland Terminal in Poland*, Tuesday 6 July 2010, http://www.maritime-executive. com/pressrelease/ hhla-and-polzug-intermodal-inaugurate-high-performance-hinterland-terminal-poland-2010-07-06/ [accessed 29 March 2011].

Horst, M.R. van der and de Langen, P.W. 2008. Coordination in Hinterland Transport Chains: A Major Challenge for the Seaport Community. *Maritime Economics and Logistics*, 10(1–2), 108–129.

SNN 2004. *Beter Goed(eren) Vervoer, Regiovisie goederenvervoer Noord-Nederland* (Regional Freight Transport Vision, in Dutch). Samenwerkingsverband Noord Nederland.

Schuttevaer 2009. *Gekoppeld varen naar Coevorden.* Weekblad Schuttevaer, 9 April 2009.

Roso, V. and Lumsden, K. 2010. A Review of Dry Ports. *Maritime Economics and Logistics*, 12(2), 196–213.

VIL 2005. *Best Practices Hinterland Connections, Strategic Working Group Hinterland Connections Ports* (in Dutch). Vlaams Instituut voor Logistiek (VIL).

Visser, J., Konings, R., Pielage, B.J. and Wiegmans, B. 2009. *A New Hinterland Transport Concept for the Port of Rotterdam: Organisational and/or Technological* Challenges?. Paper to the Transport Research Board, January 2009. Delft: TU Delft.

Chapter 4

Port Community Systems in Maritime and Rail Transport Integration: The Case of Valencia, Spain

Salvador Furió

4.1   Introduction

The advent of containers in the 1950s led to a revolution in international freight transport by giving support to a global production system where companies are migrating to seek cheaper labour costs, and supply chains are becoming increasingly complex and global. In this context, containers have been highly successful, becoming a basic element in supply chains where they are considered standard units for transport, production and distribution (Notteboom and Rodrigue 2009). The success of containers is evident if we analyse the evolution of container traffic throughput at ports, with average annual growth rates of over 10 per cent and more than 430 million TEUs in 2009, despite the 15 per cent decline experienced this year due to the global economic crisis (Drewry Shipping Consultants Ltd. 2009).

The container shipping business has focused for years on reducing maritime transport costs through economies of scale. This has led to a high level of concentration in the industry, the use of increasingly large container ships and the development of new hub-and-spoke systems where a few hub-ports concentrate the cargo. Due to the difficulty of achieving significant additional cost reductions in maritime transport, shipping companies are paying greater attention to inland transport, port–hinterland connections and integrated door-to-door services. This seems logical if we consider that inland transport costs can represent between 40 and 80 per cent of total container transport costs along the logistics chain (Notteboom and Rodrigue 2009). But this renewed interest in port–hinterland corridors is also in the spotlight for container terminal operators and port authorities. On the one hand, container terminal congestion is a widespread problem due to high container traffic growth rates and one of the possible solutions to this problem is the development of efficient corridors with regular rail shuttle services connecting the port with inland terminal extensions or dry ports (Roso et al. 2008). On the other hand, competition is no longer between ports, but between door-to-door chains. Therefore, port authority strategies include port–hinterland relations and the new concept of port "regionalisation" (Notteboom and Rodrigue 2005) has emerged, implying a more efficient maritime-land interface with an integrated corridor approach where inland terminals are directly connected to ports by rail or barge, creating real complexes or logistics networks which integrate ports, inland intermodal terminals or dry ports and logistics and distribution centres.

Dry ports are key elements of these new logistic complexes, acting as inland nodes for the concentration of goods, empty container depots and other added value logistic services. These inland hubs allow the concentration of the volumes needed for direct frequent intermodal connections with ports and other intermodal terminals. As a result, this intermodal network allows small ports to reach a larger hinterland and large ports to increase their throughput and benefit from economies of scale. However, in order to build this model, sound maritime–rail integration is necessary at physical (infrastructure), operational and information levels, ensuring dry ports are coordinated with sea ports and with all the different stakeholders that are involved in maritime–rail operations, such as maritime terminals, maritime agents, railway operators and railway undertakings. It is important to point out that coordination between all these stakeholders does not occur spontaneously. Collective actions, such as public governance by a port authority or the development of Information and Communication Technology (ICT) systems for a sector or industry, can contribute to improving coordination (Van der Horst and De Langen 2008).

ICTs have played a significant role in transport development for a long time, and the evolution of these technologies (telephone, telex, public switched telephone network, private networks, LAN, MAN, WAN, GAN networks, client/ server models, electronic data interchange (EDI), internet, etc.) comes together with the evolution of logistics and transport. (Bollo and Stumm 1998)

ICTs are crucial for the development and management of intermodal transport systems due to the dynamics of these systems and the multiplicity of actors involved. They give support to the three typical decision levels: The strategic level planning involving the design of the intermodal transport system and considering time horizons of a few years requires approximate and aggregate data; The tactical level planning basically referring to the optimisation of the flow of goods and services through a given logistics network; The operational level management or short-range planning, involving transportation scheduling of all transporters on an hour-to-hour basis, subject to the changing market conditions as well as to unforeseen transportation requests and accidents. (Dotoli et al. 2010)

This chapter's next focus is on information integration at operational level and aims to contribute to better maritime–rail integration by developing standard procedures and messages for the management of computerised information flows regarding dry port activity. This will boost efficiency in intermodal corridors connecting ports with their hinterland. First of all, a brief review is provided for the most relevant stakeholders involved in maritime–rail operations and the main operations and activities in dry ports are identified. Then, a detailed analysis is carried out on information flows for import and export operations managed through dry ports, identifying the main problems and shortfalls. This analysis and previous experience in standard messages for maritime container logistics are the basis for developing a standard message framework to computerise information flows related to dry port operations. Finally, the role that PCS plays in order to simplify the implementation of the proposed standards is analysed, and the first results of a pilot test involving the Dry Port of Madrid and the Port of Valencia are presented.

4.2   Main Stakeholders in Maritime–Rail Integration

Maritime–rail operations involve a large number of stakeholders who interact at seaports and inland terminals. Process standardisation and the advanced management of information flows will be key factors to better coordinate and enhance the efficiency of logistic chains. But before analysing these information flows, it is convenient to define the main stakeholders in maritime–rail integration that will appear later in the chapter:

- Railway Operator: It is the company that organises and offers rail transport services. It may or may not coincide with the railway company providing the rail traction.
- Railway Undertaking or Railway Company: It is the company that provides rail traction in a rail transport service. It has the necessary training, licences, certificates and authorisations to provide this rail traction service.
- Maritime Agent: It is the company that represents the shipping company at a port. Maritime agents manage everything needed by the shipping company at every call a ship makes at a port. In regular line traffic, the maritime agent also has a commercial function, looking for cargo and being responsible for container management in their local area.
- Railway Terminal: Facilities to provide train services such as train reception and expedition, train composition and decomposition, train loading and discharge, etc. When talking about maritime–rail integration, there are railway terminals at seaports connected to inland railway terminals or dry ports. Some particular aspects of railway terminals at seaports are listed below:
- They can share the yard (for container storage) with maritime terminals;
- They can be operated by the same company running the maritime terminal;
- Loading and discharge operations are sometimes performed by port stevedores.
- Depositor (at dry port or railway terminal): The company or individual that deposits the full or empty container in the railway terminal (by truck or by rail) until the container is transferred to the next transport mode (the depositor is usually the railway operator or the maritime agent for empty containers).
- Customs: Public administration responsible for safeguarding the entry and departure of goods in a country and for collecting taxes as applicable. It authorises, supervises and controls the entry and departure of goods from customs bonded areas.
- Shipper/Freight Forwarder: Shipper is the term normally used to describe the exporter or the company that requires a transport service. A freight

forwarder is an expert in international transport that normally gives support to the shipper, arranges the shipment and prepares the required documentation (freight forwarders can sometimes act as carriers).

- Road Haulier: It is the company responsible for container road haulage services. If they do not use their own means (or lorries) they are known as road transport agencies.

4.3   Dry Port Operations

The dry port concept is based on a seaport directly connected by rail (and/ or other high capacity transport means such as a barge) with inland intermodal terminals where customers can leave/pick up their units (containers) directly at/ from a seaport (Roso et al. 2010). In theory, dry ports should function as an inland extension of a seaport terminal in such a way that it becomes a new alternative for the user to pick up or deliver their container, without any additional complexity. In this sense, coordination and integration between dry port operations and the seaport is a key issue for dry port success.

Dry ports are, therefore, much more than simply inland intermodal terminals and should include additional services to give support to maritime container logistics and act as real seaport terminal extensions. Customs services, inspection services, empty container depot services and other added value logistic services such as stuffing and stripping or warehousing can be also integrated in dry ports.

Nevertheless, dry port activity revolves around the intermodal rail terminal where container interchanges from train to road and vice versa are performed. The most common activities or operations are:

- Reception and delivery of containers by lorry (i.e. gate-in/gate-out operations)
- Operations for train reception and expedition and train composition and decomposition
- Train loading and discharge
- Container yard management (full and empty container deposit and storage)

But in order to better focus the analysis of information flows on maritime–rail operations and dry ports, it is also worth learning who the clients of a dry port are (i.e. to ascertain existing commercial relations). The main clients of dry ports are normally railway undertakings and railway operators. The former contract train operations at terminals and train loading and discharge services. The latter normally deposit the containers at the dry port and pay for container deposit, gate-in/gate-out operations and other container handling services. Finally, when empty container depot services are provided, the clients who use this service are normally shipping companies or maritime agents.

4.4   Information Flows in Dry Port Operations

4.4.1   Main Problems

Nowadays information management has become essential for the development of almost any kind of activity. As we have seen before, this is even more important in dry ports due to the need for coordination with seaports and a wide range of different stakeholders. However, in most of the cases of maritime–rail operations, there is a lack of standardised procedures and a low level of implementation of new ICT tools, resulting in the use of low quality information and operative inefficiency. The main problems identified through interviews with maritime and inland intermodal terminals include:

- Lack of standardised procedures in dry port relations with different stakeholders
- Intensive use of paper
- Lack of implementation of new data exchange systems (Acklam 2007) through secure and computerised communication channels (intensive use of fax, phone and email)
- Confusion over the roles of different stakeholders in the process
- Duplication of information and information inconsistencies (Törnquist and Gustafsson 2004)
- Accumulation of errors in the information
- Insufficient information exchange of container data and lack of anticipated information which causes inadequate planning for dry port operations (Van der Horst and De Langen 2008).

4.4.2   Standard Information Flows for Import and Export Operations

The European Union is establishing the basis for the definition of standard systems giving overall support to intermodal rail freight transport (Acklam 2007). However, this is an ongoing process and the current real situation at many dry ports or intermodal terminals lacks the implementation of advanced information exchange systems with involved actors, as has been previously described. In the next few paragraphs a proposal defining standard information flows in port and dry port relations for import and export operations is defined.

The figure below represents the information flows in a multimodal import operation passing through a seaport and dry port or inland intermodal terminal (maritime–rail–road). Each column or vertical line represents one of the stakeholders while information flows are represented by horizontal arrows between them, defining the origin and destination of the information transmitted which is described briefly in a caption above the arrow.

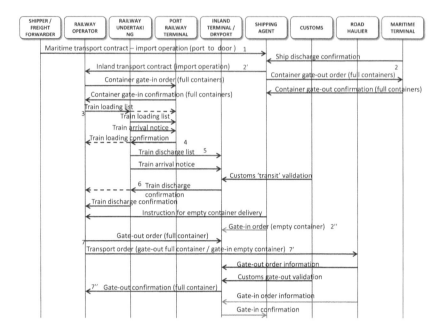

Figure 4.1   Information flows in a maritime–rail–road import operation

- *Information flow no. 1.* The shipper or freight forwarder asks a maritime agent for rail transport from the seaport indicating the dry port or intermodal railway terminal destination, the final destination of goods and the customs for clearance or transit procedures. This could be agreed previously by the shipping company contracting a door-to-door service directly and instructions being included in the Bill of Lading. In this case, information can be included in the Discharge Manifest so that the maritime terminal can position the container properly in order to avoid additional movements.
- For merchant-haulage (land leg controlled by the freight forwarder) such a request is not necessary, but the maritime agent should also be informed about the rail transport to give the proper instructions to the maritime terminal.
- *Information flow no. 2.* The maritime agent asks the railway operator for rail transport and sends a gate-out order to the maritime terminal together with a gate-in order to the empty container depot where the empty container should be placed once emptied.
- For merchant-haulage, the freight forwarder asks the rail operator for rail transport directly. Notwithstanding, the maritime agent should be informed in order to be able to send proper gate-in and gate-out orders.
- On the other hand, the railway operator should send the gate-in orders to the railway terminal in the seaport to receive the full container to be loaded on the train.

- Both the maritime terminal and the railway terminal at the seaport (they can coincide) should confirm gate-in and gate-out operations.
- *Information flow no. 3.* The railway operator sends the train loading list to the railway undertaking and the railway terminal at the seaport. This train loading list should be confirmed by the railway undertaking, which is responsible for the final loading list.
- *Information flow no. 4.* The railway terminal at the seaport loads the train and sends loading confirmation. This information is required by the railway undertaking, the railway operator, customs and the dry port or destination inland railway terminal.
- *Information flow no. 5.* Based on train load confirmation, the railway undertaking sends the destination dry port or railway terminal the discharge list including the containers to be unloaded, their position on the train (coach number and position), and the depositor of the container in the dry port (the depositor is usually the railway or intermodal operator who will give the instructions to the dry port regarding container delivery once unloaded from the train).
- *Information flow no. 6.* Once the train is received in the dry port and unloaded, the dry port (inland railway terminal) sends the unload confirmation to the railway undertaking informing them about possible incidents or damages. The railway operator also receives this information.
- *Information flow no. 7.* The railway operator (or the container depositor in the dry port) sends a gate-out order to the dry port to authorise container delivery to the road haulier for the last leg of the journey. The road haulier also receives instructions to pick up the container in the dry port, deliver the goods to their final destination and return the empty container to the empty container depot established by the maritime agent in stage 2.

The same analysis has been performed for a multimodal road–rail–maritime export operation where export goods pass through a dry port before reaching the seaport. The figure below represents the associated information flows.

Focusing on information flows to and from the dry port, two main blocks can be identified: One related to train loading and discharge operations and the other to road transport gate-out/gate-in operations. Dry port integration in the supply chain will improve as a result of standardising and computerising these information flows. This will reduce errors, improve the quality of information and increase the efficiency of dry port operations and door-to-door intermodal services.

One important step in this direction is the establishment of standard messages giving support to these train loading/discharge lists and gate-in/gate-out operations. In order to do this, it is necessary to pay attention to standard international messages already in use, such as those established by UN/CEFACT which have been adopted and implemented by the SMDG group to give support to the maritime container logistics chain (SMDG 2002). SMDG is an association involving companies and other entities involved in the maritime industry (such as container terminals,

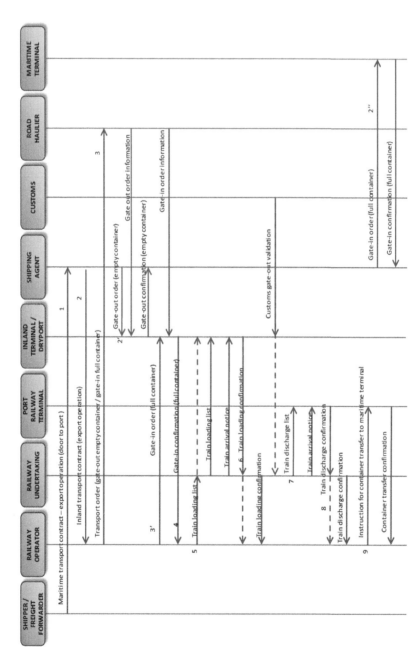

Figure 4.2   Information flows in a road–rail–maritime export operation

shipping companies or port authorities) created (among other objectives) to develop and promote a standard framework of messages to be used on maritime container logistics chains. This standard framework is a benchmark for the implementation of this kind of solution in ports and maritime terminals and could also be used as a reference when looking for similar applications in dry ports.

Table 4.1   SMDG standard messages for maritime container logistics and shipping planning

Message	Description
COARRI	Container discharge / loading confirmation
CODECO	Container delivery confirmation (gate-in/gate-out confirmation)
COEDOR	Container stock report
COHAOR	Container handling order
COPARN	Container pre-announcement and release notice
COPINO	Container pick-up notice
COPRAR	Container discharge / loading order
COREOR	Container Release order
COSTCO	Container stuffing / stripping confirmation
COSTOR	Container stuffing / stripping order
DESTIM	Container damage and repair estimation
BAPLIE	Bay plan
MOVINS	Stowage instructions

Using SMDG messages for maritime container logistics as a basis, we now present the message proposal for import and export operations at dry ports, standardising:

• Train loading and discharge orders and their confirmation
• Gate-in/gate-out orders and their confirmation

In order to implement and test the solution proposed it was necessary to readjust the messages (COPRAR and COARRI) to the specific needs of rail transport. These new versions of the messages for rail or dry port purposes are not included in this chapter.

Besides loading, discharge lists and confirmation, it would also be possible to implement other additional messages which are being used in ship operations to inform about ship bay plans and stowage instructions (BAPLIE – Bayplan, MOVINS – Stowage instructions), in case of similar information needs for train plans and train stowage instructions (currently, SMDG standard messages already include a BAPLIE for rail). This option has not been considered in the proposal due to the fact that the COPRAR and COARRI messages proposed can include information about train composition and the positioning of the containers in train coaches.

The message interchange proposal for import and export operations is presented below specifying the steps to be followed and the SMDG standard messages to be used.

4.4.3   Dry Port Message Interchange Proposal for Import Operations

1. The railway operator will send the dry port a COPRAR message containing the proposed train container discharge list. This message will include information about who is the depositor of each of the containers to be discharged and handled as well as an admission order for the container in the dry port.
2. The railway undertaking will send the dry port a COPRAR message containing the final train container discharge list (container discharge order) with information about container location in train coaches.
3. Once the train is discharged, the dry port will send both the rail operator and the rail undertaking a COARRI message confirming train container discharge and reporting differences or incidents.
4. The depositor (which is normally the railway operator) will send the dry port a COREOR message with the release order of the full import container in order to deliver the container to the road haulier performing the last transport leg.
5. Once the full container is delivered, the dry port will send the depositor a CODECO message to confirm container delivery or gate-out.
6. If necessary, the maritime agent will send a COPARN message to the dry port in order to receive and store the empty container (after being emptied in importer facilities).
7. Once the empty container is received, the dry port will send the maritime agent a CODECO message with confirmation of the empty container reception.

4.4.4   Dry Port Message Interchange Proposal for Export Operations

1. If necessary, the maritime agent will send a COPARN message to the dry port in order to deliver an empty container to a road haulier.
2. Once the empty container is delivered, the dry port will send the maritime agent a CODECO message with confirmation of the empty container delivery.
3. The railway operator will send a COPARN message to the dry port with the admission order of full containers in the dry port (this should include information in advance about the train these full containers will be loaded on).
4. The dry port will send a CODECO message to the railway operator confirming reception of each full container in the dry port.
5. The railway operator will send the dry port and the railway undertaking a COPRAR message containing the proposed train container loading list.
6. The railway undertaking will send a COPRAR message to the dry port with the final train container loading list and loading instructions.

7. Once the train is loaded, the dry port will send the railway undertaking and the railway operator a COARRI message confirming train container loading and reporting any differences or incidents if necessary.

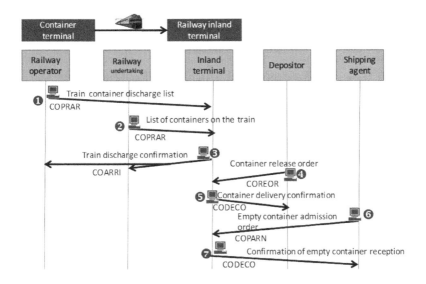

Figure 4.3   Standard messages proposal for import operations at dry ports

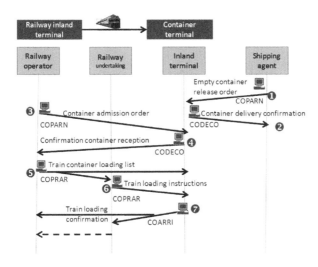

Figure 4.4   Standard messages proposal for export operations at dry ports

4.5   The Role of Port Community Systems in Maritime–Rail Integration

Port Community Systems (PCS) are technological platforms that integrate different stakeholders in port operations and maritime transport by giving support and managing information exchanges associated with the main port operations and administrative procedures. PCS originated from the use of ICT tools for the development of Single Windows to simplify transport and trade procedures by establishing a single entrance point for standardised information and documents which are requested and required by different institutional bodies (Port Authority, Master's Office, Customs, etc.). The scope of PCS differs a great deal from one port to another. Normally, they support the establishment of Single Windows for Loading, Discharge Summary Declarations, dangerous goods and port formalities (such as ship call requests and ship port clearances) but they can extend their scope both to the maritime and overland segments of the logistic chain by standardising, computerising and managing information exchanges between stakeholders for other operations along the container logistics chain (Furio and Llop 2008). PCS therefore play an important role as an intermediate platform that simplifies information flows and allows users direct access to and integration with a wide range of transport operators and public bodies for information exchanges related to transport and trade, avoiding the need to develop customised integration projects for each particular company or public entity.

This conception of PCS, which goes beyond port operations to cover a wider view of the supply or logistics chain, should play a key role in maritime–rail integration by serving also rail operations and simplifying the adoption of the previously defined standards. This will contribute to the integration and coordination of all the different stakeholders involved in maritime–rail operations and services connecting seaports to their hinterland. In order to do so, dry ports, railway operators and railway undertakings should be integrated in the PCS and new PCS services should be developed to satisfy their needs.

One example of a PCS that goes beyond port operations is the PCS at the Port of Valencia (valenciaportpcs.net), which has been used for a pilot test of the previously proposed standards for information exchanges involved in dry port operations.

Valenciaportpcs.net is a web platform resulting from the evolution of the PCS of the Port of Valencia and other EDI software applications for information exchanges between port community stakeholders. This platform has integrated more and more processes and operations, both from the maritime and overland segments, covering a longer stretch of the logistics chain and facilitating relations, communication and coordination between the different stakeholders involved. The final objective is to enhance the efficiency of the supply chain (valenciaportpcs.net 2010).

Currently more than 400 companies use the technological platform of the Port of Valencia on a daily basis, generating a yearly flow of around 25 million messages.

Some of the services provided by this technological platform include:

- Sea side: Management of documentation proceedings for booking contracts and shipping instructions prior to the Bill of Lading
- Port: Management of proceedings for ship calls, dangerous goods and loading and discharge summary declarations
- Land side: Management of lorry transport orders, gate-in/gate-out confirmation
- General track-and-trace information throughout the chain
- In order to test the framework of messages defined for rail transport and dry ports, additional land services have been developed, but are still to be commercialised by the platform.

4.6   Pilot Test in Madrid Dry Port

Madrid dry port was the first dry port project in Spain. It aimed to create adequate infrastructure and operational procedures for the development of efficient intermodal rail transport services connecting main Spanish container ports with Madrid and to therefore reinforce the competitive position of Spanish ports. This project also contributed to the consolidation of Madrid as a top-level logistic node. Today around 60,000 TEUs pass through Madrid dry port every year, most of which come from or are bound for the Port of Valencia, which accounts for around 80 per cent of Madrid dry port traffic. Some interesting characteristics of Madrid dry port are the participation scheme (integrating four different port authorities and the regional government), the railway terminal infrastructure and equipment, the development of a maritime customs enclosure in the dry port, or the development of an empty container depot. Nevertheless, there are still many things that can be done in order to improve efficiency and maritime–rail integration, such as the implementation of new and updated ICT tools and solutions.

Two pilot tests have been carried out in Madrid dry port with the support of the valenciaportpcs.net platform land-side services. In the first instance, the dry port implemented standard and computerised information exchange procedures for gate-in and gate-out orders. The second pilot involved standard and computerised information exchange procedures for train loading and discharge orders. The dry port, the maritime terminal, a railway operator and several road hauliers have participated in these pilot tests.

Different studies confirm that integrating ICT into intermodal transport systems reduces information inconsistencies, increases possibilities for effective planning (Törnquist and Gustafsson 2004) and leads to a more efficient management, in terms of system resources utilisation and overall cost index (Dotoli et al. 2010)

The pilot results at Madrid dry port confirm the benefits of implementing these kinds of solutions to improve the efficiency and control of maritime–rail operations.

With the implementation of computerised gate-in / gate-out procedures, the dry port achieved benefits such as:

- Reduced time for gate-in and gate-out operations, as the new smooth procedure is based on advanced information and computerised lorry identification by barcode readings or lorry plate readings and identification. This has reduced the time for gate controls (to less than one minute) and the total time lorries spend in the dry port (18 minutes for two operations what means more than a 20 per cent time reduction)
- Improved efficiency due to better resource planning for the reception and delivery of containers (improved efficiency is aligned with time reduction)

Concerning the implementation of automated train discharge and loading procedures, the dry port achieved benefits such as:

- Reduced train discharge and loading time and increased railway terminal capacity through the reduction of loading rail line occupation due to a better planning of train loading and discharge operations. This was possible because of the availability in advance of accurate information by implementing computerised information exchange procedures between the stakeholders involved (train loading and discharge time reductions from 10 per cent)
- Reduced administrative work at the dry port by avoiding the need to have people typing the loading and discharge lists into the terminal management system manually (administrative staff could be reduced)
- Improved dry port services sending accurate information regarding terminal operations and container location to its users

4.7   Conclusions

Maritime–rail integration is vital to progress in European transport policy towards a more sustainable and co-modal transport model and also in the maritime industry, where shipping companies, container terminals and port authorities are paying increasing attention to door-to-door corridors and port–hinterland connections. Besides the need for infrastructure integrating rail and maritime operations, information technologies can play an important role in maritime–rail integration by managing information flows between the different stakeholders efficiently and effectively. An analysis of the current situation shows a large number of shortfalls in information flow management for maritime–rail operations at seaports and dry ports, where there is a lack of standards and where the technologies used are obsolete.

A detailed analysis of the processes and associated information flows in maritime–rail import and export operations made it possible to define a standard framework for compiling and organising the main information exchanges between different stakeholders. This standard framework has been defined and presented along with a selection of standard message formats already being used in the maritime industry and which can give support to the identified information flows. Once this standard framework was defined, we looked into how Port Community Systems could integrate maritime–rail operations into their services, simplifying and fostering the adoption of these standards.

The pilot tests performed in Madrid Dry Port with the support of the Port Community System of the Port of Valencia conclude that significant benefits can be gained by implementing the general framework proposed. These benefits include service quality improvement and cost reductions (operation and administrative costs) in maritime–rail operations in dry ports and seaport terminals. Pilot results also confirm the important role that Port Community Systems can play by integrating maritime–rail operations in their technological platforms.

4.8   References

Acklam, J. 2007. *Information Interchange in Rail Freight: Improving Customer Service by Innovative use of the Telematic Applications for Freight Regulation.* Community of European Railways and Infrastructure [Online]. Available at: http://www.cer.be/media/070507_TAF_TSI.pdf [accessed 5 August 2010].

Bollo, D. and Stumm, M. 1998. Possible Changes in Logistic Chain Relationships Due to Internet Developments. *International Transactions in Operational Research,* 5(6), 427–445.

Dotoli, M., Fanti, M.P., Mangini, A.M., Stecco, G. and Ukovich, W. 2010. The Impact of ICT on Intermodal Transportation Systems: A Modelling Approach by Petri Nets. *Control Engineering Practice,* 18, 893–903.

Drewry Shipping Consultants Ltd. 2009. *Annual Container Market Review and Forecast 2009/10.* London: Drewry Shipping Consultants Ltd.

Furió, S. and Llop, M. 2008. *Las TIC y la simplificación de procesos aduaneros para una mayor eficiencia de la cadena de transporte: El caso del Puerto de Valencia.* Proceedings of the 3rd Transport International Congress, Castellón: Universitat Jaume I, 16–18 April 2008.

Notteboom, T. and Rodrigue, J.P. 2005. Port Regionalization: Towards a New Phase in Port Development. *Maritime Policy and Management,* 32, 297–313.

Notteboom, T. and Rodrigue, J.P. 2009. The Future of Containerization: Perspectives from Maritime and Inland Freight Distribution. *GeoJournal,* 74, 7–22.

Roso, V., Woxenius, J. and Lumsden, K. 2008. The Dry Port Concept: Connecting Container Seaports with the Hinterland. *Journal of Transport Geography.* In press, corrected proof.

Roso, V. and Lumsden, K. 2010. A Review of Dry Ports. *Maritime Economics and Logistics,* 12(2), 196–213

SMDG 2002. *Guidelines to Container Messages: SMDG User Group for Shipping Lines and Container Terminals* [Online]. Available at: http://www.smdg.org/ [accessed 5 August 2010].

Törnquist, J. and Gustafsson, I. 2004. Perceived Benefits of Improved Information Exchange: A Case Study on Rail and Intermodal Transports. Economic Impacts of Intelligent Transportation Systems: Innovation and Case Studies. *Research in Transportation Economics*, 8, 415–440.

Valenciaportpcs.net 2010. *Valenciaportpcs.net Port Community System* [Online]. Available at: http://www.valenciaportpcs.net/web/ValenciaPortPCSNET/d0/ d0714577-ca5d-4c8e-a560-1923ed52d2b4.pdf [accessed 5 August 2010].

Van der Horst, M. and De Langen, W. 2008. Coordination in Hinterland Transport Chains: A Major Challenge for the Seaport Community. *Maritime Economics and Logistics*, 10, 108–129.

Chapter 5

Integrating Ports and Hinterlands:
A Scottish Perspective from the Shop Floor

Gavin Roser, Kenneth Russell, Gordon Wilmsmeier and Jason Monios

5.1   Introduction

Scotland's accessibility in terms of international markets is reflected in the limited share of total Scottish unitised freight traffic coming through Scottish ports today. For market access Scotland relies heavily on maritime services via remote southern seaports, with the result that the majority of Scotland's trade travels overland through England.

Other than the Channel Tunnel, all unitised freight comes to the UK by water but this traffic accesses Scotland in different ways. Scottish ports are not attractive to deep-sea traffic due to the physical requirements of large vessels. Therefore container traffic to Scottish ports comes by feeder or short sea vessels from English or Continental European ports, carrying transhipped deep-sea containers or short sea containers originating in Europe. Scotland also has a RoRo ferry connection from Rosyth to Zeebrugge (Belgium) and from Stranraer and Cairnryan to Northern Ireland, the latter for domestic traffic. Figure 5.1 illustrates the diversion of Scottish freight flows.

Yet the evidence suggests that Scotland does not suffer poor direct maritime access with the Continent due solely to geographic or economic reasons. Lagging infrastructure development as well as a lack of sufficient government, initiatives to promote direct links have also been cited as key reasons (Baird 1997, Baird et al. 2010a).

A need to coordinate future development of port capacity, terminal operations and hinterland connections has been identified (Baird et al. 2010b).The aim of this chapter is to relate these findings to the experience on the shop floor, in order to test their relevance and match theory with practice. This chapter is based on an in-depth interview with Kenneth Russell (Roser 2011), forming the basis for a detailed case study of transport operations and hinterland strategies in Scotland. Mr Russell is a Director of John G. Russell (Transport) Ltd, a family-owned business with a history dating back to 1969. He is the second generation of the family in the business and the fourth generation in logistics, reaching back to early 1900. As Marketing Director, Mr Russell continues to spend time on the shop floor, in the warehouse and on rail sidings, monitoring train loading operations: "as a result if there is a delay I can tell a customer at first hand the cause and that is very important. So I am close to the coal face; when I bid for work, I know that the company can deliver what I sell" (Roser 2011).

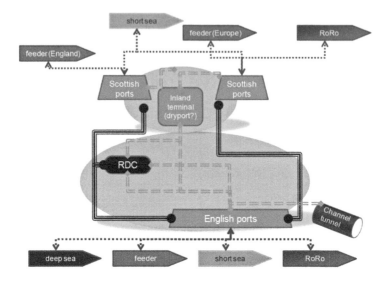

Figure 5.1   Schematic illustrating routing of Scotland's external trade
Source: Monios and Wilmsmeier (2011)

Figure 5.2   John G. Russell (Transport) Ltd rail services and terminals 2011

Russell operates three intermodal terminals (Glasgow, Inverness and Edinburgh) in Scotland and two intermodal terminals in England (Telford and Barking). Further, Russell offers rail distribution services to two destinations in Scotland (Elgin and Aberdeen) and five destinations in England and Wales (Liverpool, Cardiff, Daventry, Felixstowe and Southampton). Thus the company's rail service network provides direct services to English seaports from Scotland and accommodates the diverted trade flows to and from Scotland.

Russell does not consider that current visions and challenges for transport planning or investment are any different from those of previous generations.

> It is true today that we have much faster communications and a global economy, which raises people's expectations in terms of the services and charges they can expect from a transport company. The downside for my generation is the plethora of social, fiscal or safety regulation not to mention planning demands that we have to deal with. This can detract from the time you can devote to customers. When my father was running the business he could identify a development and, provided the financial resources were available, get on and implement his plans. Also he spent a lot more time with his customers face to face than we are able to do today. The challenges and pressures then were just as demanding but different (Roser 2011).

5.2   A Rerspective on Rail Transport Potential

In Great Britain, 85 per cent of intermodal freight train services are port-based, while 12 per cent are domestic, and the Channel Tunnel makes up the rest. Of the 12 per cent domestic, the majority of these are services between England and Scotland (Woodburn 2008a).

In terms of Scottish rail freight, Scottish deep-sea containers are moved directly to and from the ports of Felixstowe, Southampton, Tilbury and Seaforth/ Liverpool by rail by Freightliner, and DB Schenker run a service between Teesport and Mossend. Total 2009 traffic with English ports was estimated at around 73,000 TEU/year (Baird et al. 2010b). Direct container train services from UK ports to the midlands have grown over the last decade while direct services from UK ports to Scotland (i.e. Coatbridge) have fallen (Woodburn 2007). This finding represents the integration of Scottish trade flows into UK-wide distribution networks centred on key sites in the midlands and to a lesser extent north England. Woodburn (2007) noted the increasing importance of the Northwest and Yorkshire along with the Midlands for rail terminal location, and the lack of competition from coastal shipping for these inland locations. However, he questioned whether this might be altered by an increase in larger feeder vessels bringing increased traffic to other ports as opposed to the current main British ports, which would result in smaller land transport distances and a threat to the viability of rail for these flows. This potential trend could lead to the viability of port-based distribution (Mangan et al. 2008) as notions of centrality and intermediacy are altered in the UK context. Short distances in the UK have always put rail at a disadvantage with road haulage; however changes in relative distances of primary and secondary hauls due to restructuring logistics chains around port nodes could revise the calculation (Monios and Wilmsmeier 2011).

In terms of intra-UK rail flows, there is a problem with a lack of data on unitised rail freight flows. Industry figures are broken down only as far as "domestic intermodal" for the whole UK, given in tkm, which is not useful for the analysis undertaken here. However a rough estimate of 115,000 TEU for 2009 was produced by speaking to Scottish logistics companies (Baird et al. 2010b). The large DIRFT Daventry terminal (run by WH Malcolm) is the main consolidation point for Scotland, with Hams Hall also playing a role. These trains are for large supermarkets in conjunction with logistics operators such as WH Malcolm, Eddie Stobart and John G Russell, and are run primarily by Direct Rail Services and DB Schenker. Interestingly, while Coatbridge focuses primarily on port flows, much supermarket traffic from the midlands in England comes into the rail terminal at the port of Grangemouth.

According to Russell, rail is an under-utilised asset in the United Kingdom, despite the fact that "the logic of freight by rail is inescapable" (Roser 2011). The substance in that logic is evidenced by the opening of three new rail terminals from Russell between 2009 and 2011. Two of these are located in England and one in Scotland (see figure above). Russell argues that rail will succeed and has a potential to reach volumes that were carried in the 60s and early 70s due to five principal reasons:

1. Environmental factors, emissions and external costs from road congestion.
2. The UK has a good rail infrastructure, despite the lack of capital expenditure on the network in the 70s, 80s and 90s. There is also a significant secondary network which is underutilised.
3. With people still wedded to cars despite current fuel costs, their disdain for trucks continues unabated. Trucks frustrate their car journey and society does not want them in their back yards. However, the same society seems to forget that freight is a derived demand based on consumer needs.
4. The economics of scale moving critical mass by rail over road.
5. The reliability and predictability that rail services can provide both on high volume routes (e.g. Anglo Scottish) and serving regional destinations (e.g. Inverness, parts of Wales, Devon and Cornwall) where LCL (less than container load) services rather than full loads will meet the demands of Small and Medium Size Enterprises (SME) customers populating these areas. However innovative technological solutions like the TruckTrain might be required to make services in these regions sustainable. (Roser 2011)

The current main users of rail are the energy sector and aggregates industry (12 billion net tonne-kilometres in 2009/2010), while the deep sea lines (mostly moved by Freightliner) together with logistics operators like John G Russell and

WH Malcolm deliver 5.9 billion net tonne-kilometres (Department for Transport 2011). Russell suggests that untapped potential exists in the retail market and for refrigerated goods: "In the green context, retailers are only scratching at the surface. The reasons for this need to be further explored. Reliability, consistency and cost are the main drivers for retailers" (Roser 2011).

In the opinion of Mr Russell, growth has been coming in stages, but the network is currently not prepared for a fundamental modal shift, which is what is required. The network suffers from capacity limitations and gauge restrictions achieving pallet-wide viability i.e. for refrigerated units and high cube containers. To unlock the full potential of rail, it needs to be used as a tool to minimise inventory, which would require timetabled services that are both predictable and reliable. Investment in equipment such as wagons and new cranes is required to support aggregators. Further, Russell notes that "freight has the same characteristics as passengers. It needs consistency, predictability and suitable accessibility" (Roser 2011).

Other innovations to unlock the potential of rail include the "Trucktrain" concept (Bozicnik 2011). This concept builds on short trains with a carrying capacity of up to 16 TEU including configurations for reefer and high cube boxes and speeds up to 140 kph. The possibility to accelerate and break like a passenger train makes it possible to use passenger train paths, thus making it easier to schedule on the network.

5.3 Strategies for the Location of Rail Terminals

The location of rail terminals requires a clear definition of the hinterland. In his strategic perspective Russell defines "hinterland" as follows:

> A hinterland is geographic region or area serviced by any form of terminal be it a port, a hub on an inland waterway, or a terminal for rail or road. The original definition was first used in English in 1888 by George Chisholm in his handbook of 'commercial geography'. He defined it as the land behind a city or port. In Germany it historically describes the part of a country where only a few people live and the infrastructure is under developed. Actually these definitions could serve us well in the early part of the 21st century (Roser 2011).

Additionally Russell notes that the average distance of haulage journeys has an impact on where an intermodal terminal should be located. This strategy is reflected in the terminals at Hillington and Coatbridge where "75 per cent of movements are to or from destinations within 20–30 miles of the terminals, i.e. the major cities of Glasgow and Edinburgh. The remaining 25 per cent of journeys average a radius of 80–100 miles. The regional hub Inverness in a more rural region of Scotland has 20 per cent of movements within five miles, 40 per cent within 40 miles and 40 per cent within 100 miles" (Roser 2011).

Such strategic reasons were already followed for earlier developments in the company as in 1970 when John G. Russell acquired the Gartcosh (Coatbridge) site near Glasgow in 1970. That site was a) adjacent to a railway line; and b) the 37 acre site had potential for development as a container terminal, consistent with the increase in container services to the Clydeport terminal in Greenock.

Interestingly, Freightliner, the existing operator at Coatbridge, opposed the development at that time because competition was not in their plans. With the privatisation of large sections of the rail industry in 1994, different business approaches emerged and today John G. Russell Ltd and Freightliner are key strategic partners. This development represents an example of the growing trend from competition towards strategies of "co-opetition" (Song 2002) in logistics practice. 3PLs are finding that in order to achieve economies of scale, cooperation on certain routes is desirable. However industry has often been reluctant to pursue such a strategy (Van der Horst and de Langen 2008). There is also a severe inertia in the industry when it comes to location. Runhaar and van der Heijden (2005) found that over a proposed ten-year period, even a 50 per cent increase in transport costs would not make producers any more likely to relocate their production or distribution facilities. This inertia can in some ways be considered a bigger obstacle than infrastructural problems, and requires a restructuring of the transport chain in order to change transport requirements.

Figure 5.3   John G Russell intermodal terminal, Coatbridge, adjacent to the freightliner terminal

Source: Google maps

The Freightliner terminal at Coatbridge has daily rail services with Felixstowe, Tilbury, Southampton and Liverpool. There are also links from Coatbridge to the Midlands, London and Inverness. This site was a strategic development in the 1960s to provide an inland point of customs clearance for Scottish imports, therefore Coatbridge could be considered as the first Scottish "dry port".

The terminal in Hillington near Glasgow was developed under the key aspect of mainline rail access, linking to Glasgow and the West Coast mainline. This 55-acre site adjacent to the M8 and M74 provides easy access to the main North–South and East–West corridors. A further strategic asset is the location of the terminal in relation to the key retail centres at Braehead, Silverburn and Glasgow's Buchanan Galleries and St Enoch's Centre.

According to Russell, Hillington forms an ideal base for construction companies building the sports facilities and athletes' village for the Commonwealth Games 2014 in Glasgow. His vision is in stark contrast to Strathclyde Partnership for Transport who, based on a commissioned study, concluded that there was no demand for a freight consolidation centre in Hillington. The location also bears the potential to use electric vehicles for deliveries between distribution centres and retail parks. Russell stated that "so called dry port or Inland Clearance Depot (ICD) facilities and freight consolidation centres [should be seen] as one. The three critical issues for all are: location and close proximity to city centres

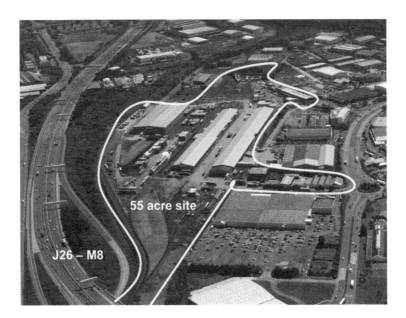

Figure 5.4   John G Russell intermodal terminal, Hillington
Source: Kenneth Russell

[e.g. Glasgow and Edinburgh], rail access, and, most importantly, access to key motorway arteries North–South and East–West" (Roser 2011).

Russell believes that customs issues are still relevant for discussions on the purpose of inland terminals, as imports originating from outside the EU still have to be customs cleared either at the seaport or the inland equivalent with customs status.

5.4   A Perspective on Planning and Investment

The development of transport infrastructure is key to maintaining the competitiveness of Scotland in international trade. Russell argues that "Scotland has made a good start with the National Planning Framework which identifies 30 key projects that are perceived as critical to competitiveness in 2030. Improvements in road connectivity to the port of Grangemouth are very welcome, the second Kincardine Bridge is very welcome and the new Forth crossing will consolidate our critical North–South corridor from Aberdeen to Edinburgh and the south. New port priorities have also been identified on the Forth which is welcome particularly since so much of our trade is with Continental Europe, therefore we must improve our sea connections. Diageo is further developing its site in Leven in Fife which will certainly add traffic volumes to the east coast road network. There are many good things going on and additional land bridge opportunities through Loch Ryan and perhaps Troon to Ireland will help us to integrate traffic flows to service a population of over 6.7 million" (Roser 2011).

However on a more critical note Russell states that "frustrations remain with regard to government regulation, at both UK and EU levels. Sometimes new regulations appear or old rules are changed without apparent reason, and short term government thinking limits the potential for growth." Russell is critical of "the traditional five-year cycle rather than long term strategic thinking in the context of investment in infrastructure, like France which takes transport strategy out of the political arena" (Roser 2011).

While a national strategy for planning infrastructure investment is laudable, questions have been raised regarding the ability of the Scottish government to fund all of these projects, and the ability to attract the private sector to drive such developments both financially and strategically will be extremely difficult (Wilmsmeier et al. 2011; Monios and Wilmsmeier 2011). A number of studies have been performed over the years for the Scottish government, providing data on freight flows and potential port and inland terminal locations or development strategies (e.g. MDS Transmodal 2002, WSP 2006, Scott Wilson 2009). Yet, despite locations being promoted in government policy and planning, it has been difficult to develop a strategy whereby both government agencies and private stakeholders can achieve maximum benefit for minimum risk. In evaluating such complex situations, an appreciation of the political and institutional relations are required.

Key priorities to be addressed by the industry are more effective coordination between the UK Treasury, the Department for Transport (DfT) in London and the European Commission in Brussels to maximise renewal and upgrading of rail infrastructure. "Without effective coordination we have no hope of fully utilising the existing network to its potential. I have already mentioned the secondary network which is grossly under used. I must stress this is against a background where governments always place passengers first over freight" (Roser 2011).

Russell feels that the principle of a European rail network without borders is undeniable: "National rail jurisdictions just do not make sense. We as the UK need to closely examine our potential future domestic network and how it links to the European network" (Roser 2011). In Russell's understanding transnational cooperation and networking in groups such as the European Freight and Logistics Leaders Forum are decisive in Brussels as these allow for sharing best practice with shippers and transport suppliers in other EU countries. The fact that he is one of only two UK members shows the significant need for UK operators to engage in and develop pan-European strategies.

5.5 Key Priorities for Rail Development

Due to reasons of historical development, the loading gauge[1] on the UK rail network is more constrained than in other EU countries (Woodburn 2008b). Despite a range of upgrades over recent years, gauge continues to be a significant challenge. Despite the fact that "increasing from W8 to W10[2] in many cases is just a matter of paperwork, the DfT does not even look at changes when minimal cost is involved" (Roser 2011). These constraints represent a specific challenge for intermodal transport of deep sea containers, as many routes linking ports to inland terminals have not progressed beyond the W8 loading gauge (max height 8'6" on standard rail wagons) that was implemented to all major ports as a consequence of the maritime container revolution. Moreover, high cube (9'6" height) containers are expected to increase to 65–70 per cent of the market by 2023 (Network Rail 2007). In order to facilitate transport of these units the enhanced loading gauge (W10) or low ride specialist wagons (e.g. Barber Low Ride 14.25) are required. Purchase and maintenance of specialist wagons is typically more expensive and they reduce

---

1 "The physical dimensions of a railway vehicle and its load are governed by a series of height and width profiles, known as loading gauges. These are applied to a given route to ensure that a railway vehicle will not collide with a lineside or overline structure such as platforms, overbridges or tunnels. Loading gauge profiles vary by route, reflecting the constraints on vehicle size caused by lineside and overhead structures" (Network Rail website (a)).

2 W8: Allows standard 2.6 m (8 ft 6 in) high shipping containers to be carried on standard wagons. W9: Allows 2.9 m (9 ft 6 in) high Hi-Cube shipping containers to be carried on "Megafret" wagons which have lower deck height with reduced capacity. At 2.6 m (8 ft 6 in) wide it allows for 2.5 m (8 ft 2 in) wide Euro shipping containers which are designed to carry Euro-pallets efficiently (Network Rail website (b)).

the available train payload, thus making them economically undesirable for freight operators (Woodburn 2008b, Network Rail 2007).

In Scotland Inverness, Aberdeen and Fort William are all gauge-constrained and there is a need to consider alternative solutions to costly infrastructure investment. "A wagon solution would be far more cost effective," says Russell (Roser 2011). However government assistance is likely to be required for this solution to be implemented. According to Russell "in the case of the gauge problem the responsible government departments are not aware of the benefits than can be achieved for the economy in excess of perceived savings through cuts" (Roser 2011).

Woodburn (2006) suggested the use of government grant to aid the start-up of "flows that are not immediately commercially viable but which are likely to become so within a reasonable period. This would overcome the existing 'chicken and egg' situation. With a funded trial, viability could be established and further traffic that could use the new service could be identified." While this idea is promising in principle, it is likely that the government would struggle to placate competing private companies if grant were given to one company to run a speculative service and build up its business with the commercial risk transferred to the public sector. In addition, Monios (2010) found that only a small proportion of the annual Scottish modal shift funding budget has been spent each year (e.g. £3.7m spent out of £15.4m in 2008/9). Reasons found for this lack of spending include the lack of strategic identification of projects, the lack of centralised knowledge and responsibility, and the misalignment between funding requirements and eligibility that results in difficulties attracting bids for the money. Now that the annual budget has been reduced to £2m (after the initial decision to scrap the grant was reconsidered due to industry pressure), many operators have suggested that they intend to bid for this money.

One of the questions that arises is whether transport companies really look consciously at the overall economy and its ups and downs when planning investment priorities over medium term time windows. If a company like John G. Russell is building a new terminal, buying new cranes or setting up a new route, how can they sure that the business will be there? Russell says that it is unrealistic to expect shippers to sign long term contracts due to their inability to predict the future actions of the economy or their customers.

> What we can do however is anticipate trends in key sectors, look at demographics and the resultant critical mass and the impact of government legislation, particularly in the context of rail as road congestion at its present levels and growing is not sustainable. On the specifics of retailers, people will continue to eat and consume multiple products, so we conclude that if one customer today supplies these products and they are not there tomorrow then someone else will fill the breach. That is how the economy works. Having said that, the transport industry needs to respond rapidly to changes in the economy and the supply chain demands of key shippers; we never take that for granted (Roser 2011).

Every company has four stakeholders: customers, shareholders, employees and suppliers. Therefore the key must be to deliver commitment through consistency. If the offering is based on green credentials then it should be sustainable: "Companies that do not adopt this creed will struggle with delivery of service and credibility – this certainly applies to many elements of rail" (Roser 2011).

"Innovation is key to survival", according to Russell, "and if companies are innovative and looking to the future they are more likely to survive any issues that arise." Moreover, companies must "build rail into an end-to-end solution. Generally customers want a seamless offering." But the question is whether rail is genuinely attractive to the retail sector? "It can be with appropriate planning, including working with the customer and Network Rail to unlock the paths that provide the service required, but it does not fit all" (Roser 2011). Research has also shown that a service needs to be well-developed before shippers will use it (Van Schijndel and Dinwoodie 2000). This problem has been encountered in Scotland with the only international ferry service for Scottish shippers, between Rosyth and Zeebrugge, Belgium. Changing schedules and even a complete break in service when the original operator withdrew from the market in 2008 have contributed to Scottish hauliers driving down to England to access ferry services to the Continent.

Before the advent of computer systems, RFID and bar codes, information management was all done manually, however Russell noted that a mistake on the computer system can lead to delayed customer payment. This is a significant issue in an industry where "the gross margin in a good year ranges from 3 per cent to 5 per cent" (Roser 2011). Cash flow is therefore imperative to survival. Russell says that "one major challenge we must face as a company and a country is the imperative not to lose sight of the commercial imperative to innovate and generate cash. As an industry we must make money and generate profit margins that are sufficient and sustainable in order that we are still around to invest in the future" (Roser 2011).

Thinking from the customer perspective is very important for a logistics provider, according to Russell.

> Cost neutral – then it is interesting. Cost effective – then it is needed. But you must not compromise service; if it is more expensive – forget it. Furthermore, simply switching to rail is insufficient to achieve carbon savings; the train capacity use remains crucially important. To attain environmentally friendly results the minimum capacity use on the train has to be achieved. Sending half empty trains can be positively negative in fulfilling environmental benefits (Roser 2011).

In Russell's perspective rail needs to be a credible product in its own right; not just because it is presumed to be environmentally friendly: "Shippers need detailed information on emissions and external costs so that when decisions are made to use rail they are made for the right reasons" (Roser 2011).

In terms of new sites for Scotland, Russell suggests that there is a case for a site located on the west coast at Fort William. Otherwise he is not convinced based on current evidence that any new sites are needed. However development is required on the current network. Scotland's key intermodal terminal at Coatbridge needs improvement: new cranes and a new layout would allow a significant increase in throughput. Additionally, siding capacity and length is in need of improvement at Coatbridge, Mossend and Grangemouth. "If this could be improved it would speed up train turnaround times, improving track occupation time. This would result in more capacity capability" (Roser 2011).

A number of key operational areas also need to be improved, according to Russell: higher travel speeds, more acceleration and deceleration to aid in pathing and improved end to end times; a well-disciplined timetable, predictability; train traceability, always aware of performance; higher productivity in fuel workforce and train assets.

5.6   The Shop-floor Perspective on Future Developments in Scotland

The integration of ports and hinterlands from a Scottish perspective is influenced by two primary infrastructural issues: the lower gauge on the East Coast Main Line (which is used to divert trains when the WCML (West Coast Main Line) is unavailable) and the problems with high-cube (9ft6) containers north of the central belt. With high cubes expected to account for the majority of 40ft containers in the near future, this problem needs to be addressed. As the cost is too prohibitive to raise the required bridges to allow high cubes through, the only feasible option is to use low wagons. However there is a lack of these wagons in the UK, and they attract higher maintenance costs. Moreover, under current regulations government funding cannot be used to solve this problem. Additionally, even when funding is available for a new rail service, the length of lead time can be detrimental to the service development process.

Access has been reduced as the UK rail industry has seen a major decline in wagonload services over the last few decades. Better information for potential shippers is also required regarding train services, timetables and wagon capacity. Due to a lack of marketing and information availability, rail is often not visible to prospective customers. According to Russell it is necessary "to utilise capacity far better. We see far too many less than filled trains running around. We need a trading platform that shares the resource available" (Roser 2011).

Public sector initiatives can help to resolve this issue through feasibility studies and knowledge of other experiences to bring a new service or facility to fruition by reducing the risk that either the shipper or the operator has to take in the first instance. In the meantime, rail services will rely on large shippers and then smaller users can add their containers to these regular shuttles.

Furthermore, Russell clearly highlights the importance of aligning service offerings with customer needs, as well as maintaining reliability and consistency.

It needs to be at the right price for both supplier and customer or it will not be sustainable. It needs to satisfy all. These requirements come back to innovation, which was mentioned earlier. Innovative ways to enhance capacity on the network need to be pursued, in particular gauge enhancements on specific routes and wagon solutions where required. Where possible, faster, cleaner, more fuel-efficient trains need to be implemented (Roser 2011).

The current situation has significant implications for the practicalities of conducting trade to and from Scotland as a peripheral location. Major adverse impacts have been observed from the current arrangements as the hinterland transport of Scottish cargo and underlying logistics structures have not been given the necessary relevance in policy and by the private sector, where a visionary mid- and long-term perspective remains uncommon. Therefore the shop floor perspective provided in this chapter has confirmed previous findings in academic research presented earlier. Industry, government and academia can work together to develop innovative solutions to the problems of infrastructure upgrading and collaboration on the provision of capacity on intermodal services. A research agenda can be developed from this perspective that incorporates not just a supply side approach (focused on infrastructure and services) but the demand side as well, bringing together large and small shippers to underpin economically viable services on key transport corridors.

5.6   Acknowledgements

Research for this chapter was undertaken with the financial support of the EU-part-funded Interreg IVb North Sea Region project Dryport.

5.7   References

Baird, A. 1997. A Scottish East Coast European Ferry Service: Review of the Issues. *Journal of Transport Geography.* 5(4). 291–302.

Baird, A., Grieco, M. and Wilmsmeier, G. 2010a. *Overcoming Territorial Discontinuity: The Need for an Evolutionary Approach for Ferry Services Provision in Scotland.* Paper presented at the annual conference of the American Association of Geographers, Washington, April 2010.

Baird, A., Monios, J., Wilmsmeier, G. and Mathie, I. 2010b. *The Effect of Unitised Freight Flows and Logistics Strategies on Scotland's External Trade.* European Transport Conference, Glasgow, October 2010.

Bozicnik 2011. Interdisciplinary Solutions for the New Railway Freight System [Online]. Available at: http://www.fpp.edu/~mdavid/TVP/Seminarske%2008-09/ICTS2005CD/papers/Bozicnik.pdf [accessed May 2011].

DfT 2011. *Transport Statistics Great Britain.* Department for Transport.

Mangan, J., Lalwani, C. and Fynes, B. 2008. Port-centric Logistics. *The International Journal of Logistics Management*, 19(1), 29–41.

MDS Transmodal Ltd 2002. *Opportunities for Developing Sustainable Freight Facilities in Scotland*. Report prepared for the Scottish Executive, Edinburgh.

Monios, J. 2010. The Effect of Maritime Policy and Funding on Short Haul Shipping in Scotland. Paper presented at WCTR, Lisbon, Portugal, July 2010.

Monios, J. and Wilmsmeier, G. 2011. Dry Ports, Port-centric Logistics and Offshore Logistics Hubs: Strategies to Overcome Double Peripherality? *Maritime Policy and Management*, forthcoming.

Network Rail 2007. *Freight Route Utilisation Strategy*. Network Rail.

Network Rail website (a) [Online]. Available at: http://www.networkrail. co.uk/browse%20documents/rus%20documents/route%20utilisation%20 strategies/great%20western/great%20western%20rus%20baseline%20 information/03.%20infrastructure/loading%20gauge/loading%20gauge.pdf [accessed 29 May 2011].

Network Rail website (b) [Online]. Available at: http://www.networkrail. co.uk/browse%20documents/rus%20documents/route%20utilisation%20 strategies/great%20western/great%20western%20rus%20baseline%20 information/03.%20infrastructure/loading%20gauge/gauge%20diagram.pdf [accessed 29 May 2011].

Roser, G. 2011. Interview with Kenneth Russell. 15 April 2011. Unpublished.

Runhaar, H. and van der Heijden, R. 2005. Public Policy Intervention in Freight Transport Costs: Effects on Printed Media Logistics in the Netherlands. *Transport Policy*. 12(1), 35–46.

Wilson, S. 2009. *Scottish Multi-Modal Freight Locations Study*. Edinburgh: Scott Wilson.

Song, D.-W. 2002. Regional Container Port Competition and Co-operation: The Case of Hong Kong and South China. *Journal of Transport Geography,* 10(2), 99–110.

Van der Horst, M.R. and De Langen, P.W. 2008. Coordination in Hinterland Transport Chains: A Major Challenge for the Seaport Community. *Maritime Economics and Logistics*, 10(1–2), 108–129.

Van Schijndel, W.J. and Dinwoodie, J. 2000. Congestion and Multimodal Transport: A Survey of Cargo Transport Operators in the Netherlands. *Transport Policy,* 7(4), 231–241.

Wilmsmeier, G., Monios, J. and Lambert, B. 2011. The Directional Development of Intermodal Freight Corridors in Relation to Inland Terminals. *Journal of Transport Geography*, forthcoming.

Woodburn, A. 2006. The Non-bulk Market for Rail Freight in Great Britain. *Journal of Transport Geography*, 14(4), 299–308.

Woodburn, A. 2007. The Role for Rail in Port-based Container Freight Flows in Britain. *Maritime Policy and Management,* 34(4), 311–330.

Woodburn, A. 2008a. Intermodal Rail Freight in Britain: A Terminal Problem? *Planning, Practice and Research.* 23(3), 441–460.

Woodburn, Allan G. 2008b. The Challenge of High Cube ISO Containers for British Rail Freight Operations, in *Logistics Research Network Annual Conference 2008: Supply Chain Innovations – People, Practice and Performance*, 10–12 Sep 2008, University of Liverpool, UK (unpublished).

WSP 2006. *Scottish Freight Strategy Scoping Study*. Report prepared for the Scottish Executive, Edinburgh.

PART II
Africa

Chapter 6

Dry Ports and Trade Logistics in Africa

Charles Kunaka

6.1   Introduction

Efficient logistics performance is a fundamental element of trade and economic development, and particularly so in a global economy that is widely interconnected and interdependent. Reducing logistics costs is critical to the ability of developing countries, particularly those in Africa, to participate more in international trade. Remoteness which is partly a function of trade and transport logistics can exacerbate the isolation of countries and hamper their participation in global production networks. There is considerable evidence that trade costs across Sub-Saharan Africa are much higher than in other parts of the world. MacKellar, Wörgötter, and Wörz (2002) established that transport prices for most African landlocked countries range from 15 to 20 per cent of import costs, figures that are three to four times more than in most developed countries. In fact, Amjadi and Yeats (1995) established that transport costs in Africa have become a higher trade barrier than import tariffs and trade restrictions. Limao and Venables (1999) explore this aspect further and argue that the costs of trade are important determinants of a country's ability to participate fully in the world economy.

The relatively poor logistics performance of countries in Africa is apparent when one looks at the Logistics Performance Index (LPI) developed by the World Bank. The LPI offers insights into the logistics performance of countries across the world. It covers the entire supply chain and is based on a survey of logistics professionals worldwide. Coordinating the various stages of product development, component production, and final assembly requires the ability to move goods across borders quickly, reliably, and at low cost (Arvis, Mustra, Ojala, Sheppard, and Saslavsky 2010). The LPI can help identify reform priorities within countries. It is based on numerical ratings of 1 (weakest) to 5 (strongest) to assess logistics performance.[1] The overall performance of African countries in 2010, the latest available index, is given in Figure 6.1.

---

1   The International LPI is based on the assessment of foreign operators located in the country's major trading partners, and is a weighted average of six components: 1. Efficiency of the customs clearance process; 2.Quality of trade and transport-related infrastructure; 3. Ease of arranging competitively priced shipments; 4. Competence and quality of logistics services; 5. Ability to track and trace consignments; and 6. Frequency with which shipments reach the consignee within the scheduled or expected time.

In the 2010 LPI the average score for countries in Sub-Saharan Africa was 2.42 which were the lowest of all regions of the world. However, The LPI shows distinct variation in the logistics performance of individual countries in Africa, though the majority is at the lower end of the global ranking. South Africa is the best performing country, ranked 28th in the world with a score of 3.46, while Eritrea (154th) and Somalia (155rd) are the worst performers amongst the countries surveyed, with scores of 1.70 and 1.34 respectively. All the other countries fall in between, though generally in the lower third of the list f countries listed by rank. Overall the LPI points to weaknesses in all aspects of logistics performance but most pronounced in customs, infrastructure and quality of logistics services. Typically, where it concerns international trade, overall performance of a supply chain will depend on the performance of the weakest link in the chain.

Unless countries in Africa are able to reduce their trade costs, then they will be excluded from global supply chains that have become so common under the trend towards trade in tasks. In order to determine measures to bring countries further into the global trading system it is important to understand both the determinants of trade costs, and the magnitude of the barriers to trade that the costs create. The barriers can be multifaceted, including the quality of both infrastructure and logistics services. The latter dimension is explored by Teravaninthorn and Raballand (2008) who study transport corridors in different parts of Africa and find that the degree of competition in transport services has a major impact on transport

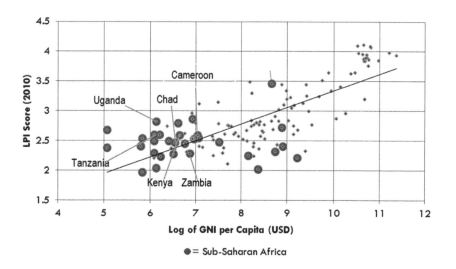

● = Sub-Saharan Africa

Figure 6.1   Logistics Performance Index, 2010
Source: Own calculations, LPI data from Arvis, et al. (2010)

prices. Even when transport costs are not that much higher than in developed countries, the contestability of the transport markets impacts significantly on the prices that are faced. They find that across Sub-Saharan Africa, Southern Africa followed by East Africa have relatively lower prices while West and Central Africa where market entry is controlled by trucker associations face much higher prices. In general, Africa's transport prices are high compared to the value of the goods exported which are predominantly high bulk and low value.

The problems faced with components of logistics systems in Sub-Saharan Africa contribute to making trade route operations unreliable. Poor reliability of logistics systems is a major problem for traders, probably more so than the average transit time. It increases logistics costs as shippers have to carry more inventories, suffer stock-outs and disruptions to operations, make emergency shipments at higher costs and lose markets. Trade corridors have many components and if each is characterised by poor predictability of performance then the whole system because overly unreliable. Figure 6.2 shows the time in days it took to clear containers at the Port of Mombasa, Kenya in 2004. Half the containers going to Uganda and Rwanda from the port of Mombasa were cleared for transit within nine days, but one in 20 took more than a month. Within this context dry ports will have a positive effect on operations if they are able to make corridor systems more reliable otherwise if they are also unreliable then they will add to the problems that are faced.

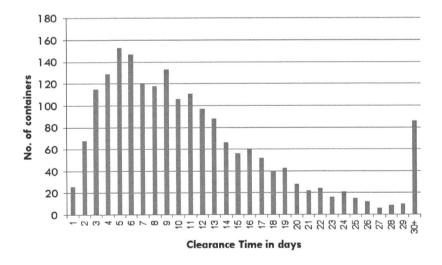

Figure 6.2   Distribution of clearance times at Port of Mombasa
Source: Arvis, J.F., Raballand, G. and Marteau, J.F. (2007)

6.2   Dry Ports in Trade Logistics

A general approach to improving trade logistics in Africa has centred on trade and transport corridors. A "trade corridor" is a multi-modal trade route connecting points of economic activity along its length. Its primary focus is on economic efficiency and it should ideally provide users with transport choices. A corridor approach offers a holistic planning framework, covering hard and soft infrastructure and general institutional development to improve the trade facilitation environment. It is in the context of corridor projects that dry ports have typically been developed.

There are numerous definitions of dry ports across the literature, including those offered by the UN agencies and others by the OECD. Suffice to say for purposes of this assessment, the operational definition of a dry port is a an inland terminal which is directly linked to a seaport and which offers cargo clearing and handling facilities similar to those available in a seaport. This functional definition is deliberate to include the different names that facilities similar to dry ports are called in different countries. Two of the more common names are inland container depots (ICDs) and container freight stations (CFSs). Obviously the two have their specific definitions and not all ICDs and CFSs are dry ports.

Arvis, Carruthers, Smith and Willoughby (2011) maintain that ICDs have evolved as a convenient intermediate solution between clearance of cargo at the border, which is generally the least convenient option for shippers, and clearance on the buyer's premises, the most convenient option for the importer, but least convenient for customs. They are often located in the outskirts of a hub city where the price of land is moderate and arterial highways and railways give good access and do not interfere with urban traffic.

The above suggests that there is a potential optimal location and function of a dry port. An appropriate location would be one which balances the needs of the different parties that are involved, shippers of goods, control agencies and transport and logistics service providers. Roso, Woxenius and Olandersson (2006) provide a classification of dry ports based on their location. They distinguish between distant, mid-range and close dry ports.

Distant dry ports are located at distances from sea ports to exploit the comparative advantage of rail and river transport. These modes are ideal for moving large volumes over long distances. Under this scenario, dry ports help consolidate traffic which then benefits railway operations in particular. In fact, in several places, for instance in Tanzania, the dry ports were developed, owned and operated by the railways. Storage and customs and other border management operations were available on site. Generally, distant dry ports are the most common type of dry port that is found in Africa. Mid-range dry ports are at intermediate locations from seaports. They help facilitate the consolidation of traffic by road transport which is then still moved by rail to and from the seaport. Close dry pots are in close proximity of seaports and mainly serve to increase the space available for port operations. Several of the new generation CFSs, especially in East Africa, is in the immediate neighbourhood of the major seaports. They often have their own customs and border management services on site.

While this typology helps to understand where dry ports are located and the function they serve, it is important to acknowledge also that the geographical layout of corridors or countries typically also influences the development of dry ports. Below we use several examples to show how local issues, including historical evolution of transport networks, influence the design of dry port systems in Sub-Saharan Africa.

The rest of this paper explores the role that dry ports play in the trade logistics systems in different parts of Africa. Five case studies are used to identify locational factors, connecting transport systems, functions and ownership and management attributes of the dry ports. The first case study is of a dry port in Tanzania which was designed to serve Malawi; the second is on the largest dry port in Africa, which is in South Africa; the third case study explores how dry ports have been used to address port congestion in Tanzania and the last two case studies focus on a new generation of dry ports, which are focused much more on border management processes in Niger and Ethiopia. All the case studies are on major trade corridors in Africa.

6.3 Experience with Dry Ports in Africa

There is a long history of dry ports in Africa going back some four decades. Most were developed in the 1970s and 1980s. Africa has a large number of landlocked countries. Not surprisingly, several of the dry ports were designed to serve landlocked countries or the inland regions of coastal countries. In order to deliver goods to the interior in best possible conditions of speed and security, multimodal terminals were set up within the network, which are used as "forward ports" for pre- and post-transport operations. The facilities were equipped with infrastructure, equipment and systems to receive and process cargo. They were and still are a considered a useful tool in the development and trade integration.

There have been two main phases of development of dry ports across Africa. The first phase was in the 1970s to 1990s when dry ports were developed as part of railway networks in particular. The second phase started at the turn of the millennium and coincided with the thrust towards private sector management of ports and railways as well as the general drive towards integrated logistics systems. Already new dry ports are being developed or are proposed in Ethiopia, Niger, guinea, Rwanda, Uganda and several other countries. There has therefore been a marked change in ownership of dry ports with some developed by shipping lines, others by port operators and yet others by third party logistics service providers.

6.3.1 *First Generation of Dry Ports*

Up until the 1990s railways across Africa were often state owned monopolies. Most were constructed at the turn of the 20th century and had thrived during the mineral booms up to the middle of the century. However, nearly all started to experience declining traffic volumes especially in the 1970s. This was also the period when road transport started growing, with the increased adoption of containers. Dry ports, or more accurately ICDs, were developed as a way of maintaining railway competitive edge, especially given the long distances involved in moving international trade traffic. Examples of dry ports developed around this period were City Deep in South Africa and Mbeya in Tanzania. However, the two were developed for completely different reasons.

6.3.1.1 City Deep, South Africa

The largest dry port operation in Africa is City Deep which is located in Johannesburg, South Africa. City Deep was developed in 1977 by South African railways. It is connected by road and rail to the port of Durban, which is the largest container terminal in Africa. It has also hinterland connections by road and rail to several landlocked countries to the north: Botswana, Zimbabwe, Malawi, Zambia and Democratic Republic of Congo. The port is equipped with rail mounted gantry cranes and reaches stackers. It has more than 2,000 terminal ground slots for import and export, more than 700 slots for empties but has no reefer slots.

The Durban to Johannesburg route (part of the North–South Corridor in Southern Africa) is the busiest in Africa, handling more than 20m tonnes of cargo per year, representing more than two thirds of all import–export containers in South Africa. The Johannesburg region is home to numerous manufacturing and logistics enterprises. In 2003 some 40 per cent of general cargo moving on the route passed through the dry port facility. The dry port is designed to receive 50 wagon container trains on a schedule. In 2009 there were up to 19 trains per day between City Deep and Durban. However, the average for the past several years has been five trains per day, of which three trains went to the dry port terminal and the other two went to private sidings. Presently annual throughput is approximately 220,000 TEU (50 per cent full imports, 30 per cent full exports, 20 per cent empty exports). The capacity of the dry port is estimated at 375,000 TEU per annum (Portfutures 2003). Trains take 16 to 18 hours' travel time to cover the 600km distance between Durban and Johannesburg, while the turnaround time for trains is five to eight days.

Though the traffic to the dry port is moved by rail, the railway share of traffic between the seaport and Johannesburg is only about 30 per cent. This is mainly because the seaport to dry port railway operation is plagued by several problems which until recently were causing the railway to lose market share. Some of the main problems that have been faced over the years are:

- Inadequate and unreliable capacity. The railways are not able to provide adequate capacity and most of the new growth in traffic has been carried by road transport. Some cargo moved by road still passes through the dry port if it has been selected by customs in Durban for physical inspection.
- Delays in the dry port. Trains coming into the dry port are first stopped in

the marshalling yards for checking and splitting which could take several days. In addition, transferring shorter trains to sidings also contributed to long turnaround times for trains.

- Change of locomotives. While the Durban–Johannesburg track is electrified line, the dry port is not and there has to be a change to diesel locomotives, which also contributes to delays.
- Poor security. The railways have also been previously affected by poor security leading to pilferage of goods in transit. However, this is has to a large extent been addressed by improving security around the dry port.
- Lack of space to handle long truck configurations. The dry port does not have enough space to handle the "superlink" truck configuration which is favoured by trucking companies in the region (Portfutures 2003).

Indications are that the decline in market share has been arrested and traffic volumes have stabilised. There are now plans to build a new larger terminal some 35km to the east of City Deep.

While there are these problems on the South Africa domestic movement, the problems are worse on the international traffic movements from City Deep. Some cargo destined to the landlocked countries is picked up or delivered either by truck or by rail from/to City Deep. The problems that are faced in both cases are most apparent at the border where significant delays are experienced. Figure 6.3 below shows the amount of time it takes to cross the two most important border posts on the corridor from South Africa to Zimbabwe, Zambia, and DRC. Curtis (2009) attributes these problems to the long border processes and clearance times.

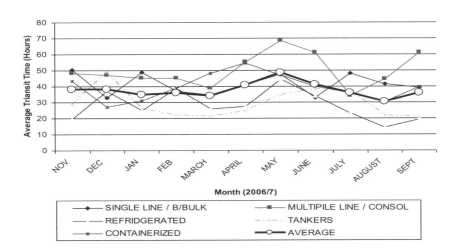

Figure 6.3 Chirundu border post crossing times, Zimbabwe into Zambia, 2006/7

Source: Curtis (2009)

The problems are no better when it comes to the railways. Figure 6.4 shows the time–distance graph for a shipment moving by rail from Durban to Lubumbashi in the DRC. As is apparent, there are significant delays at each border crossing or handover point (Beitbridge, Victoria Falls, Ndola). This is largely due to the need to change locomotives. Though the railways of the different countries are interconnected, only wagons can move across borders and locomotives have to be changed. This process is prone to considerable delays, reducing the economic speed of the railways to a walking speed.

In the past, the railway operators in Southern Africa have tried to develop through freight train services running across different countries. The system worked well up to the mid-2000s when some of the railways were concessioned or fell into disrepair. Subsequently, the services collapsed as either the concessioned railway operators were not keen to participate or turnaround times in the poor parts of the network became too unpredictable. Equipment could therefore be held up for long periods with negative knock-on effects across the regional network. Individual operators therefore sought to minimise risk by restricting locomotive operations to their domestic network and also charging demurrage fees for wagons that are not returned on time.

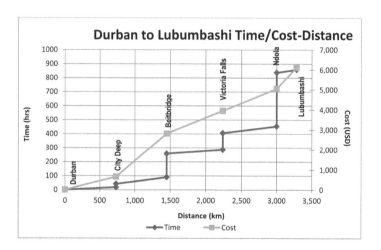

Figure 6.4   Time and cost distance for railway operations on the North–South Corridor, 2008

Source: Own estimates, data from various sources

6.3.1.2   Mbeya, Tanzania

Malawi is a landlocked country in Southern Africa. For access to the sea it has traditionally relied on ports in neighbouring Mozambique. In fact, the railway from Blantyre the commercial hub of Malawi to the ports of Beira (640km) and

Nacala (815km) in Mozambique handled over 90 per cent of Malawi's external trade. During the 1980s, a civil war was in full swing in Mozambique, disrupting trade routes with neighbouring countries. By December 1983, the line to Beira was closed and in July 1985 the line to Nacala closed as well. Malawi therefore needed an alternative trade route, urgently. The available alternative trade route was much longer. This was the route from Malawi to the port of Durban in South Africa via Zambia and Zimbabwe (a distance of 3,500km from Blantyre). Estimates are that in 1984 total losses to the economy from using this alternative route were equivalent to 20 per cent of the value of all exports. This was a huge burden on the economy. The alternative was to connect Malawi to the Port of Dar es Salaam in Tanzania, a distance 1,600km from Blantyre. There were already in place a road and railway line in Tanzania running from Dar es Salaam to the Copperbelt of Zambia and to the Democratic Republic of Congo. Both passed just over 100km north of the Malawi border with Tanzania. This outlet was identified as Malawi's only reliable trading link with international markets.

In order to operationalise this route several investment were made, with donor support. The main one was the construction of a dry port at Mbeya in Tanzania to facilitate transshipment of cargo between road and railway systems. Mbeya is 105km from the Malawi border. The facility was designed to allow Malawi cargo to travel by rail between the seaport and Mbeya (a distance of 880km) and then be transferred to road trucks. Other components of the project were improvements of the road between Mbeya and the main centres in Malawi, construction of fuel transshipment facilities at Mbeya; provision of fuel railway tankers and ordinary wagons; and construction of a border post and weighbridges. The whole system was designed to provide a secure alternative at lower cost as long as the more direct alternatives remained closed. Management of cargo passing through the corridor was handled by a public private enterprise called Malawi Cargo Centers (MCC). MCC was awarded space also in the Port of Dar es Salaam where it could handle Malawi destined cargo.

Obviously, the project was high risk as the duration of the closure of the low cost routes was not known. Still, it was a success such that by 1993, the Dar es Salaam route carried close to 20 per cent of Malawi's external trade, and was particularly important for fuel imports which increased from one per cent in 1986 to around 40 per cent of all fuel imports by 1994. Even today, there are significant volumes of fuel and smaller volumes of other products that are shipped through the alternative trade route. However today, the Mbeya dry port facility handles very little traffic. Peace returned to Mozambique in 1992 which allowed the Nacala rail link and the Beira road link through Tete to be reopened, and the substantial investments into the MLS and MCC were essentially marginalised. Independence of Zimbabwe and majority rule in South Africa in 1994 eliminated the political reasons to avoid the Southern routes as well.

Despite the reopening of the Southern routes, the corridor through the dry port could still be attractive but only if the railway line was performing well. However, the railway services passing through the dry port are slow and unreliable. It currently handles only about 12 per cent of its design capacity. The apparent competition between trade routes shows clearly that services passing through a dry port should offer a clear logistic advantage for them to be used by shippers.

One of the other current operational problems faced is that the Tanzania customs office at Mbeya where acquittals for traffic in transit are processed is not computerised and therefore not linked to the main network of Tanzania customs. As a result, there are delays in handling acquittals and releasing customs bonds for transit traffic passing through the dry port. The decline in cargo volumes also places huge demands on MCC to market the route and to demonstrate the costs saving possibilities, based on a logistics business cost rationale.

6.3.2   New Generation of Dry Ports

After years of road, rail and port projects being developed in isolation, a far more strategic and coherent approach towards logistics is now being adopted in Africa. Increasingly port developments are more explicitly linked to improvements to land transport systems. This is the case particularly for the movement of bulk cargo; but is also applicable to containerised cargo. This more holistic approach should pay dividends in the longer term, boosting not only the continent's ability to export to the rest of the world but also increasing trade between neighbouring African states. As part of this new approach, shipping lines, port operators, and third party service providers are showing interest in managing the land-side logistics facilities and services. An example of a dry port being developed by each of these categories is described below.

6.3.2.1   Mojo, Ethiopia

Ethiopia is a landlocked country in Horn of Africa. Although Ethiopia has seven potential sea outlets there are only four that are active. The major import and export corridor is to Djibouti. The other ports are either too far such as Mombasa and Port Sudan, not accessible due to political problems (Asab and Massawa in Eritrea) or in unstable territory or on poor condition (Berbera and Mogadishu in Somalia). As a result, the corridor that runs from Addis Ababa to the Port of Djibouti (Ethio–Djibouti Corridor) handles about 97 per cent of Ethiopia's foreign trade. The balance is primarily trade from Northern Ethiopia through Port Sudan. The Ethio–Djibouti Corridor consists of three components: the port of Djibouti, the inland transportation and the dry ports. The Port of Djibouti is managed by DP World and is one of the more efficient ports on the east coast of Africa.

Ethiopia and Djibouti are mutual dependent on each other in that for Djibouti the port is an important revenue generator while for Ethiopia the port is the only feasible outlet to the sea. More than 80 per cent of the traffic volume through the Port of Djibouti is coming from or going to Ethiopia. The transport links between the two countries go back to the last decade of the nineteenth century, when a decision was taken in Ethiopia to construct a railway line between Addis Ababa and Djibouti. Despite financial and political problems, the Addis Ababa–Djibouti railway was completed in 1917. The completion of the railway line proved decisive factor in making Djibouti the main outlet for Ethiopian trade.

Following the construction of the railway line, the redirection of traffic to Djibouti was rapid. By 1925, export volumes transiting through Djibouti were four times the volumes in 1910. The main railway had lower transport costs compared to the caravan trade. The railway dominated traffic until the 1950s when competition with road transport increased. Between 1953 and 1957 rail traffic fell by 50 per cent a trend that continues today when the road transport carries over 90 per cent of the traffic between Djibouti and Addis Ababa. Overall, the links through Djibouti continued to handle Ethiopian traffic until the Ogaden war of 1977–78. Between 1977 and 1998 usage of the Corridor by Ethiopia generally collapsed. Traffic was redirected to Assab at the onset of the war. The railway line was damaged and traffic ceased. A change of fortunes occurred in 1998 as traffic to Djibouti recovered when the Ethiopia–Eritrea war broke out. Since then Djibouti has once again become the main outlet for Ethiopia. In 2009 the Ethiopia traffic handled at the port was 124,000 TEU of imports, 33,000 TEU export and 91,000 TEU empty exports. A significant proportion of cargo containers are de-stuffed in the port, largely due to the high demurrage charges levied by shipping lines. The charges are a reflection of the perceived poor turn round time for containers travelling to Ethiopia from the port.

The two countries have a transit agreement which provides the legal framework for cooperation on the corridor. Ethiopia is using the agreement as the basis for its dry ports strategy. The government has designed a plan to develop a network of dry ports and freight stations throughout the country. The dry ports will serve as centres designed to reduce logistics costs by allowing consolidation of traffic in regional centres, greater convenience for shippers as customs processing will be close to final destination, and promote use of multi-modal transport. The proposed system will be a departure from current practice where all customs processes are carried out in the capital, Addis Ababa.

The first dry port under this plan opened in 2010 at a place called Mojo, some 35km from Addis Ababa on the Ethio–Djibouti Corridor. The facility is intended to be the primary and principal dry port of the country around which a national network will be developed (Ahmed 2010). The location of the facility is close to the existing highway and railway to Djibouti. However, the railway is presently not operating fully though there are plans to rehabilitate it. Based on current practice, cargo arriving in Djibouti is processed for transit on site by Ethiopia customs which have an office in the port. The cargo is then moved to Mojo where processing for final clearance is performed. The dry port is expected to lead to quick removal of cargo from port thereby reducing port related storage costs, and port congestion as well as serving as a collection and consolidation facility for cargo. Figure 6.5 shows the time cost distance graph for cargo passing through the Mojo dry port.

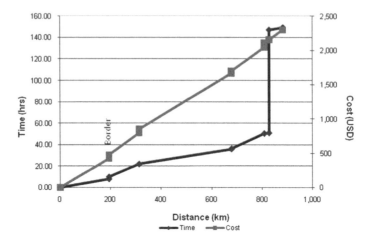

Figure 6.5   Time and cost distance for operations on the Ethio–Djibouti Corridor, 2010

Source: Own estimates, data from various sources

The new dry port is managed by the government owned Dry port State Enterprise Corporation. However, operations between the seaport and the dry port are handled by the Ethiopia Shipping Lines. Trade traffic to and from Ethiopia has by law to be handled by the shipping line which as a monopoly. However, there are indications that its rates can be as much as 50 per cent higher than other shipping lines. This practice can therefore impose a huge cost on the economy.

6.3.2.2   Dosso, Niger

One of the new dry port projects about to be developed is in Niger. Niger is a landlocked country connected by four main trade corridors to seaports. The dominant corridor is the one to Cotonou in Benin. It is estimated that this corridor currently carries some 40 per cent of Niger's overseas trade traffic. The corridor is 1036km long from Niamey to Cotonou. It includes the port, a road and a 438km long railway line, from the port to Parakou, in Benin. In Cotonou, shipments are typically sent by train to Parakou, where they are then loaded onto trucks. However, the railway line has very limited capacity, at most 200,000t per year and is also facing increasing competition from road transport. As a result, road trucking is very important on the corridor. But the road sector also faces numerous challenges including high prices due to the *tour-de-role* practices followed by operators. This has numerous intertwined consequences including low truck utilisation rates (around 40,000km per year (Teravaninthorn and Raballand, (2008)), an old and inefficient truck fleet and vehicle overloading. Operators therefore prefer to de-stuff containers at Parakou so they can carry more cargo.

In an effort to improve the performance of the corridor logistics system, Niger has proposed to develop a dry port at Dosso, some 136km from Niamey on the corridor to Cotonou. Dosso is 462km from the rail head at Parakou. The dry port is proposed as a greenfield development at the junction of the main routes to Cotonou and also to Zinder, Niger's second largest city. The stated logistics related objectives of the dry port are to: a) facilitate and process international trade for Niger and promote value-added services as goods move through the supply chain; b) speed the flow of cargo between the port and major land transportation networks; c) move the time-consuming sorting and processing of merchandise inland, away from the congested the port. It is intended that the dry port would be full customs clearance and processing centre.

The dry port is to be developed as a private public partnership. Reportedly some major shipping lines have expressed interest in managing the dry port. This would be consistent with evidence elsewhere that facilities that are managed as an integral component of a system that includes a port and connecting transport system perform better than those that are standalone.

While the objectives of the dry port are consistent with typical role of such facilities, it is not apparent that the proposed location is the most ideal. Within Niger itself there are two other locations that have been considered. The first is Niamey, the capital. Under the current plan cargo would be transferred from rail to road trucks at Parakou and handled again at Dosso, away from its final destination which may increase costs. According to a 1999 study some 80 per cent of the cargo moving along the corridor is destined for Niamey, which may make it an ideal location. In fact under current plans, there remains a proposal to develop another dry port at Niamey, possibly under the same management as the one at Dosso. The other possible location is at the border town of Gaya. Gaya is at the junction of the corridors to Cotonou and Nigeria though the border post at Kamba. To add to the challenge of selecting an ideal location, the dry port is also being proposed in a region where there are numerous other dry port projects. Similar facilities are either operational or planned in Northern Nigeria, Burkina Faso and also in Mali. After all, some 80 per cent of the cargo is destined there. A regional approach may therefore be warranted, especially given the limited traffic volumes flowing across this part of the world.

The railway line has been proposed for concessioning which may help to address the perennial underinvestment in the system. It is not apparent what value addition will take place at Parakou, besides being a convenient customs clearance point. There has been discussion to extend the railway to Niamey, which would overcome the necessity for increased handling. Clearly a road transport only connected dry port loses some of the advantages inherent in increased scale offered by rail transport. However, in the short term the traffic volume is too small and largely unidirectional with more imports than volumes. Road transport, despite its problems, is therefore better suited to the trade volumes.

6.3.3   Container Freight Stations in Tanzania

The port of Dar es Salaam experienced a serious problem of congestion in 2007–2008, reflected in long container dwell times (Figure 6.6). The congestion was due to several problems but mainly due to the rapid increase in volumes with a static capacity or even a deteriorating ability to increase the capacity of the port's container-handling facilities. Container volumes of 370,000 TEU by 2008 were already way in excess of nominal capacity of 250,000 TEU per annum. Although the area available for storage is clearly a constraint, the capacity of the terminal was reduced further due to container dwell times for inbound containers in excess of 20 days versus five to ten days for similar ports. This led to serious congestion in the harbour, in the port and also in the surrounding access roads to the port. The congestion therefore, affected not only the downstream movement of imports, especially transit cargo for neighbouring countries, but also had an impact on upstream activities. The shipping lines, in order to limit the impact of the queue at Dar es Salaam, attempted to transship containers via the port of Mombasa in neighbouring Kenya. The congestion in Dar es Salaam not only caused a dramatic reduction in the frequency of vessel calls but also lead to a backup of deliveries with containers destined for Tanzania remaining in the transshipment ports in the middle east and in Singapore for extended periods of time.

The container terminal operator, port regulator and government and shippers came together to decide on a solution to the problem. A port congestion committee was organised to look into this and other causes for the congestion. The adopted solution involved a combination of initiatives by the terminal operator to improve the facility and procure additional equipment, by the port to provide more land, by the regulatory agency to adjust the storage tariffs, and by the private sector to develop a network of off-dock container yards, and Customs to introduce electronic submission of declarations and pre-arrival processing of declarations.

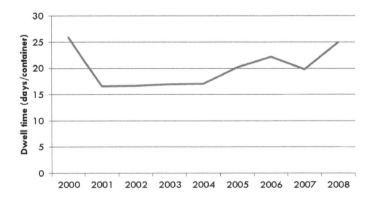

Figure 6.6   Port of Dar es Salaam container dwell time 2000–2008
Source: Data from Tanzania Ports Authority

It was also coincidental that these measures were implemented when there was a decrease in traffic as a result of the global financial crisis.

The development of the network of ICDs occurred relatively rapidly in response to the perceived profitability of these facilities. The ICDs have a total of 3,400 ground slots and therefore almost doubled the container capacity of the port. The port area has 3,600 ground slots. As initially conceived, the ICDs operated independently. Although they were nominally subject to the port tariff, there was little regulation of their pricing.

The decision on which container should be transferred to the ICDs and the procedure for allocating containers to different ICDs was initially made by the terminal operator and the port authority. Transfers were organised on a shipload basis, subject to two restrictions. Transit containers could not be moved to the ICDs and the consignee could designate a specific ICD on the bill of lading. The decision as to which vessel's containers would be transferred was made by the port authority and terminal operator in discussion with the shipping line. The ICDs to which the containers were transferred were determined through a daily poll of the ICDs to determine if they had sufficient space to receive that full load. The transfer was arranged by the terminal operator, who contracted for the trucking services. The shipping line was charged for the transfer and, in turn, collected this from the consignee.

Unfortunately this approach alienated consignees who had to pay for the transfer and the shipping lines, which had their boxes scattered among different ICDs. After a year of trial and error, a system was introduced in which there is no charge for the transfer of containers and the shipping lines select the ICD. The ICD collects the late clearance fee which is sufficient to cover the movement. The shipping lines negotiate volume discounts with the operators. The ICDs compete with the terminal by offering better service in terms of tracking containers and arranging for customs and related services.

The use of ICD was considered to be a short term approach to the problem of congestion in the port. The port authority has sought a long-term solution involving the development of a dry port, outside the port city area but not too far. A feasibility study is currently underway to identify an ideal site for the dry port.

However, it is important to recognise that a site outside the port city area is only one solution. In the case of Dar es Salaam, there are also opportunities to reduce customs clearance times and to improve land connectivity. The Tanzania customs has been improving its procedures, increasing the use of electronic submission of documents and introducing risk management procedures. With continued effort, these improvements could reduce the average dwell time of containers by three to four days. The actual reduction would be even greater since the increased certainty on the time to clear cargo would allow better scheduling of cargo movements saving an additional one to three days. Increasing the berth throughput through the introduction of more cranes and better management of the container yard would allow vessels to reduce their turnaround time. This would allow shipping lines to introduce larger vessels and to offer day of the week service. These and other efforts would increase capacity and thus allow more time for the development of new port facilities. It would also allow for the development of a freight corridor plan that would provide adequate capacity on the landside.

6.4   Function of Dry Ports in Africa

The foregoing points to several functions that dry ports are playing in logistics across Sub-Saharan Africa. The main characteristics and functions are summarised in Table 6.1.

Table 6.1   Characteristics of selected dry ports in Sub-Saharan Africa

Dry port	Country	Year established	Management	Main functions
City Deep	South Africa	1977	Railway operator	Transshipment Customs Clearance Consolidation and distribution
Mbeya	Tanzania	1986	Railway operator and public private body	Transshipment Customs transit processing
Dosso	Niger	To be developed	Shipping line	Customs clearance
Mojo	Ethiopia	2010	Government enterprise/shipping line	Customs clearance and later transshipment
CFSs	Tanzania	2008	Private operators	Customs clearance Storage

There are four main functions that dry ports in Africa:

a.   Transfer of cargo between road and rail transport, mainly to exploit the economies of scale available through use of rail transport for long distances. The first generation of dry ports was tied to the railway systems. Dry ports located in major demand generators were likely to have customs and other border management functions. In most instances the dry ports were part and parcel of the railway network. Such dry ports have tended to decline as the railways have deteriorated. This is the fate for example that has befallen the dry ports at Mbeya and Isaka in Tanzania that were developed as part of the Tanzanian railway systems. However, in recent times some of these facilities have been revived as the railways are once again receiving attention. Railway concessions in particular have enabled the private sector operators to invest in the dry ports, but only as part of strategies to enhance overall system performance. One of the best examples of this integrated approach to logistics companies with operations in several African markets are the operations of the Bollore Group. The group has operations in 41 countries. In West Africa it has a combined handling capacity of more than 1.5m TEUs per year, including termini at Abidjan in Cote d'Ivoire and Tema in Ghana. In Cameroun Bollore operates a dry port at a place called Ngaoundere where they move cargo from the port of Douala to the dry port for onward transportation by road to Chad. The operation is emerging as one of the most significant in Sub-Saharan Africa. Though government owned, the South Africa system is going through a similar revival process where increased volumes through City Deep are projected to grow in the years ahead as part of strategy to rebalance traffic between road and rail transport.

b. Customs and other border management: typically as part of both of the above, dry ports play a key role in facilitating customs and border management procedures close to the main centres of demand. The availability of customs and border management services is one of the most important distinguishing features of dry ports. In some countries it is in fact the customs authorities who licence dry ports. For example, in Rwanda a Bollore subsidiary has recently been awarded a licence from the Rwanda Revenue Authority (RRA) to operate the country's first privately owned dry port. This is a significant step in the liberalisation of the Rwandan logistics markets, where the publicly owned dry port called Magasins Generaux du Rwanda (Magerwa) has had a monopoly on bonded warehousing through a network of four ICDs. The new facility is designed to handle traffic between the ports of Mombasa in Kenya and Dar es Salaam in Tanzania on the one hand, and Rwanda, Burundi and eastern Democratic Republic of Congo on the other. About 80 per cent of all cargo moving around the East Africa region is containerised. The new dry port should also be in a good position to take advantage of the planned railway between Kigali and Tanzania's Central Railway, which will allow containers to be moved from Rwanda to Dar es Salaam by rail. The customs regime under which a dry port operates has potentially significant impacts on the efficiency of the whole system. There are several options that are available partly depending on the location of a dry port relative to the seaport. For example, the dry port could be the final destination stated on a bill of lading, or operations could be under domestic transit regime between the seaport and dry port. Other options are also possible, at times based on the level of trust between the customs authorities, dry port operator or connecting transport operator. Generally, railway operators of dry ports tend to be favoured because of the perceived level of risk that cargo could be diverted before being cleared.

c. Extension of hinterland of shipping lines. The new generation of dry ports in particular, is being developed or managed by shipping lines or third party logistics services providers. This is already the case in Ethiopia where the dry port at Mojo is managed by Ethiopia Shipping Lines, a government owned enterprise. Several shipping lines have also expressed interest in developing the dry port in Niger. Such new facilities need not necessarily be linked by rail to the seaport; they could be connected only by road transport. Shipping lines and logistics services providers are interested in dry ports because it enables them to better manage the logistics chain between seaport and inland destination, which is the most problematic leg in logistics across Sub-Saharan Africa. As competition between corridors increases, it enables them to also capture the market high up the chain. They can the arrange movement from the seaport inland and at the same time provide a facility for shippers to execute all import processes. For outbound cargo, the dry ports provide a facility of consolidation of traffic. As discussed above, there is a huge traffic imbalance in Africa where there are traditionally more imports than exports or where exports are bulk commodities which are transported using vehicles that are different from those required for import movements.

d. Consolidation of traffic: One important factor that contributes to high costs of logistics in Africa is the lack of economies of scale. Importers and exporters tend to be small in size and generate small quantities at a time. It therefore becomes difficult to justify increasing system capacity unless there is a way to facilitate consolidation of volumes so as to lower unit costs. One of the major functions of dry ports therefore is to consolidate volumes. Often, within the dry ports containers can be stripped and their contents delivered to multiple destinations, or even broken down, processed, and repackaged for multiple final buyers. Such stripping, or de-stuffing, is very common on some trade routes in Africa, especially where there are a lot of small scale traders, who ship less-than-container load volumes. Related to the above, dry ports also help reduce the amount of empty running that is experienced in both rail and road transport. In almost all countries in Africa, imports substantially exceed exports, especially in truck-borne and containerised cargo. Freight forwarders can organise traffic for trucks or containers that otherwise would have to return to the seaport empty. This can help bring shippers and truckers together through the provision and dissemination of market information about available capacity and prevailing prices. Dry ports can facilitate the market by offering facilities, offices and communications, to brokers and freight forwarders.

The review of dry ports in Sub-Saharan Africa indicates that though these facilities play an important role, they are still faced by numerous problems. Some of the issues that need to be tackled are discussed below.

6.5   Making Dry Ports Work

One of the major challenges faced with the dry ports in Sub-Saharan Africa is that the facilities have not all been developed with a logistics cost-minimisation imperative. Several were developed as discrete locations along a trade corridor, with little effort to optimise how they connect to seaports and also to the markets that they serve. The critical test from a trade competitiveness perspective is the extent to which a dry port impacts logistics costs along a trade corridor. In theory dry ports should play a facilitative role in trade logistics. They should:

- Enable fast and efficient port and dry port connectivity, if possible supported by fast and efficient railway services and with appropriate intermodal interfaces;
- Fast clearance of cargo in and out of seaports;
- Have interconnected customs and border management systems to allow fast clearance of cargo;
- Be part of a corridor length system design that seeks to minimise overall costs; and
- Be based on long term licensing and stable regulatory regimes.

These are all areas where logistics performance in Sub-Saharan Africa is weak. Consequently, it is not surprising that often cargo passing through dry ports is faced with longer delays than that which is shipped directly. Some of the main constraints that should be addressed in order to reduce costs are identified below.

6.5.1   Reduce Cargo Dwell Times in Seaports

The slow movement of containers and cargo through ports in Sub-Saharan Africa has emerged as one of the major constraints faced in international logistics. Data from numerous ports show that cargo stays in ports for periods of time ranging from five days to several months. AICD (2009) established that container dwell times in East and West Africa are 12–15 days, twice the international best practice of seven days. The problem is so serious that it is often cited by enterprises as one of the most pressing infrastructure constraints that they face. Most delays are due to long processing and administration times and poor handling in congested port areas, rather than on any real limitations in basic quay capacity.

Port efficiency is affected also by the poor performance of the connecting land transport systems. This is often exacerbated by congestion in the immediate urban areas as large cities have grown around ports. Due to the various problems, in some ports the tendency is to de-stuff containers in the port area, further contributing to delays. Containerisation is therefore not fully exploited along trade corridors, leading to the loss of the benefits of fully integrated multi-modal transport. As a result, there is little containerised traffic on some corridors serving landlocked countries. These patterns have serious implications on the performance of dry ports. Firstly, cargo passing through dry ports is subject to the long processing and clearance times in ports. The assumption of fast removal to a dry port is therefore not always correct. This is the reason why, as mentioned above, it has been observed that cargo passing through dry ports in fact faces longer transit times that that which is shipped directly from seaport to final destination.

6.5.2 *Reduce Clearing Times at Land Border Crossings*

Borders across Africa are notorious for the delays that are experienced by trade traffic. As road infrastructure has been improved, the delays faced at border crossing points have become more prominent. It is not surprising for cargo to spend four to five days or longer at each border that is crossed. The delays increase trade costs, reduce transport equipment utilisation and compromise the trade competitiveness of African countries. It is critical that fundamental reforms are executed in how cargo is cleared through border posts. This requires both infrastructure improvements but even more critically, reforms to procedures and systems. Efforts are now underway to introduce one-stop border posts in an effort to reduce clearance times. An example of this is the introduction in 2009 of a one-stop border post at Chirundu, between Zambia and Zimbabwe. However, this initiative has to be complemented by simplification and reform of the clearance process. Related to this, another measure that seems to be having the most impact is the interlinking customs and border management systems of neighbouring countries. This has been done at the Malaba border post between Kenya and Uganda resulting in the reduction of average clearance time from five days to about three hours. Without such reforms, further investments in roads and other infrastructure will have little impact on overall transit times. Moving some border management processes from the border to an inland facility should greatly reduce clearance times.

6.5.3 *Improve Railway Performance*

It is apparent that maximum benefits could be derived from dry ports if they are connected to seaports by efficient and reliable railways services. A major issue on the international corridors is that locomotives from one country generally are not allowed to travel on another country's network. As a result, rail freight crossing borders must wait to be picked up by a different locomotive. As shown in the case of the corridor in Southern Africa, delays at borders can be extensive. This is in part due to the poor availability of locomotives in the different countries, poor coordination between railway operators and absence of clear contractual incentives to service traffic from a neighbouring country's network. Reducing such delays would require a total rethink of contractual relationships and access rights linking the railways along the corridor. It would also likely require a regional clearing house to ensure transparency and fairness in reciprocal track access rights. The SADC region has in the past explored this concept, under the framework of operating through freight trains. However, the idea broke down, ironically because some of the new private sector operators were not keen to cooperate with railways in neighbouring countries or wanted to route traffic in ways that maximised their individual revenues.

6.5.4   *Liberalize Road Transport Markets*

There are a few instances where dry ports are being designed based on road transport providing connectivity to the seaport. The idea is that the dry port would provide a facility for the consolidation and deconsolidation of cargo, as well as the other usual functions including customs clearance. However, for the systems to make an impact the road transport markets have to be performing well. Liberalisation of transport markets, if possible even at regional level should allow cargo to be moved a lowest possible cost (Raballand, Giersing and Kunaka 2008).

In addition to making sure the markets are operating well, it is important to tackle also the various impediments to movement that are common along trade corridors. In West Africa, UEMOA and ECOWAS have collaborated with the USAID in collecting information on the number of barriers to movement along regional corridors. They found numerous police and customs checkpoints, some official and others informal. The checkpoints, which are not unusual trade corridors across Sub-Saharan Africa contribute to delays, uncertainty and increase the cost of transport. They have a debilitating effect on the performance of the regional transport systems so their removal is important.

6.5.5   *Facilitate More Private Sector Investment and Operation*

Some of the strategic inventions in logistics management in Africa are increasingly driven by private sector enterprises. This is true of supermarket chains trading in several countries or third party logistics service providers providing integrated logistics solutions. While international lending agencies can promote dry ports, they are best developed by their potential operators and users. The public sector can take a role in providing land and building connecting infrastructure but investment and operation of dry ports can be left to the private sector. The government also has an important role in providing appropriate regulatory frameworks that allow trade contracts to designate the dry port as a final destination. At least in several countries existing customs laws already allow this to happen.

6.6   Conclusion

It was argued in the introduction that Africa suffers from high trade costs. This is one of the most fundamental problems that are now faced in the continent's trade competitiveness. Ultimately, dry ports should be assessed based on the extent to which they contribute to reducing these trade costs. The new generation of dry ports being proposed across Sub-Saharan Africa would generally appear to be driven by logistics performance imperatives. This is important to reducing trade costs. Fortunately, designing dry ports with this objective will yield at the same time other positive results, including reducing negative environmental externalities of trade and transport systems.

6.7   Disclaimer

This paper is a product of the staff of the International Bank for Reconstruction and Development/The World Bank. The findings, interpretations, and conclusions expressed in this paper do not necessarily reflect the views of the Executive Directors of the World Bank or the governments they represent.

6.8   References

Ahmed, Y. 2010. *Final Report of Feasibility Study of Dry Port*, Report for African Trade Policy Center (ATPC). Addis Ababa: UNECA.

AICD. 2009. *Africa's Infrastructure: A Time for Transformation*. Washington, DC: AICD, World Bank.

Amjadi, A., and Yeats, A.J. 1995. Have Transport Costs Contributed to the Relative Decline of Sub-Saharan African Exports?, *Policy Research Working Paper No. 1559*. Washington, DC: World Bank.

Arvis, J.F., Raballand, G.F.R. and Marteau, J.F. 2007. The Cost of Being Landlocked: Logistics Costs and Supply Chain Reliability, *World Bank Policy Research Working Paper No. 4258*. Washington, DC: World Bank.

Arvis, J.F., Carruthers, R., Smith, G. and Willoughby, C. 2011. *Connecting Landlocked Developing Countries to Markets: Trade Corridors in the 21st Century*. Washington, DC: World Bank.

Arvis, J.F., Raballand, G. and Marteau, J.F. 2010. *The Cost of Being Landlocked: Logistics Costs and Supply Chain Reliability*. Washington, DC: World Bank.

Arvis, J.F., Mustra, M.A. Ojala, L., Sheppard, B. and Saslavsky, D. 2010. *Connecting to Compete – Trade Logistics in the Global Economy: The Logistics Performance Index and its Indicators*. Washington, DC: World Bank.

CSIR, Stellenbosch University and Imperial Logistics 2010. *Sixth Annual State of Logistics Survey for South Africa 2009*. Pretoria: CSIR.

Curtis, B. 2009. The Chirundu Border Post: Detailed Monitoring of Transit Times, *SSATP Discussion Paper No. 10*. Washington, DC: SSATP/World Bank.

Elygis, Frillet Societe d'Avocats and MTBS 2010. *Projet de port sec au Niger: Rapport de Structuration de Transaction*. Niamey: Republique du Niger.

Government of Niger 2010. *Rapport Diagnostic Final: Actualisation de la Strategie Nationale des Transports*. Niamey: Republique du Niger.

IBI Group 2006. *Inland Container Terminals: Final Report*. British Columbia.

Korea Maritime Institute 2007. *Logistics Sector Developments: Planning Models for Enterprises and Logistics Clusters*. Bangkok: UNESCAP.

Kunaka, C. 2011. *Logistics in Lagging Regions: Overcoming Local Barriers to Global Connectivity*. World Bank Study Report. Washington, DC: World Bank.

Limao, N., and Venables, A.J. 2001. Infrastructure, Geographical Disadvantage and Transport Costs. *World Bank Economic Review* 15(3), 451–79.

MacKellar, L., Wörgötter, A. and Wörz, J. 2002. Economic Growth of Landlocked Countries, in *Ökonomie in Theorie und Praxis*, edited by Chaloupek, G., Guger, A., Nowotny, E. and Schwödiauer,G., Berlin: Springer, 213–26.

Portfutures. 2003. *Johannesburg City Deep Baseline Report for Stage One*. Johannesburg: Portfutures UK.

Raballand, G.F.R., Kunaka, C. and Giersing, B. 2008. The Impact of Regional Liberalization and Harmonization in Road Transport Services: A Focus on Zambia and Lessons for Landlocked Countries, *World Bank Policy Research Working Paper No. 4482*. Washington, DC: World Bank.

Roso, V., Woxenius, J. and Olandersson, G. 2006. *Organisation of Swedish Dry Port Terminals*. Goteborg: Chalmers University of Technology.

World Bank 2008. *Niger: Modernizing Trade during a Mining Boom – Diagnostic Trade Integration Study for the Integrated Framework Program*. Washington, DC: World Bank.

PART III
Asia

Chapter 7

Dry Port:
The India Experience and What the Future Holds – India Needs to Think Out-of-the-Box

Raghu Dayal

7.1   Introduction

Intermodal terminal and transport infrastructure development in India has helped bring about qualitative changes in the realm of production as well as logistics. Dry ports, generally known in the country as inland container depots (ICDs) and container freight stations (CFSs), have made a tangible contribution to growth, both directly, through reduced transaction costs, and indirectly, through more efficient organization of manufacturing and distribution. A number of initiatives and innovations in concepts and practices have led to the growth of multimodal transport infrastructure. Even so, inadequacies, especially in intermodal transport along the arterial corridors, emphasise a dire need for capacity-building, as well as redressing distortions.

7.2   Transforming the World Economy

Changing Global Economic Geography

Transportation offerings need to grow rapidly around the world. Continuing globalisation coupled with high growth rates of population density and GDP in some regions implies a continued increase of the flow of goods and people, thereby altering the world's economic geography. From 2020 onwards, the E7 (China, India, Brazil, Russia, Indonesia, Mexico and Turkey) are expected to break away from the G7; the combined E7 GDP is projected to be around 30 per cent higher than total G7 GDP by 2030 (Pricewaterhouse Coopers 2010).

Worldwide, transport growth has been consistently higher than economic growth. World trade continues to grow, and more and more of it is containerised. Over 70 per cent of the global general cargo volumes generated are shipped in containerised form.

The Container Revolution

The container or the "box" has lent a new dimension to logistics. Just-in-time (JIT) manufacturing and precision in logistics is unimaginable without the container.

"The container made shipping cheap and by doing so changed the shape of the world economy" (Levinson 2006). As container shipping became intermodal, with a seamless movement of containers among ships and trucks and trains, goods could move from Asian factories directly to the shelves and stockrooms of retail stores in America or Europe.

New Division of Labour

With the growing incidence of outsourcing and offshore manufacturing, the market for containerisable cargo in intermodal transport has changed radically in recent years. By the year 2020, 80 per cent of the goods in the world will be manufactured in a country different from where they are consumed compared with 20 per cent now (McKinsey and Company 2010.).

Enhancing Value and Burgeoning Volumes

The dynamic growth in the container trade has mainly resulted from:

- increasing exchange of goods in the course of the growing integration of national economies and stronger international division of labour;
- increasing share of manufacturing, and value-added products in trade;
- movement of production facilities to overseas locations;
- reduction in transportation costs for containers and consequent increase in the suitability of containerisation for lower value exports; and
- continuing increase in cargo deliveries to the large seaports by means of feeder vessels.

7.3   Dry Ports Relieve Spatial Imbalances

Dry Ports Drive Growth in Hinterlands

Geography adds dramatically to the development challenges. In some cases, long distances within individual countries create vast hinterland areas such as in China, India, United States or the Russian Federation. Inland sites, economically linked to the coastal networks, ICDs and CFSs as dry ports in generic terms serve as nodes for consolidation and distribution of goods much like seaports, providing integrated multimodal door-to-door services.

A CFS, as in India, is generally an off-dock facility close to the servicing port, helping to decongest the port by shifting cargo and customs-related activities.

These are also set up inland to connect with a regional rail-linked ICD by road. Those near the ports serve a dual purpose: as an extended arm of the port, and for handling export–import cargo for coastal cities which themselves are important industrial and commercial hubs. Viewed by some as close-by dry ports, such CFSs, situated in the immediate vicinity of a seaport, provide a buffer to the port by enlarging its terminal capacity. In some cases, rail shuttle services link the CFSs with the seaport. The mid-range dry ports, usually with road-based connectivity, act as consolidation points for block train services.

Economic growth and trade in most countries has historically centred around seaports with a focus on the development of port terminals and maritime shipping networks. Coastal areas worldwide have generally been richer than inland sites and have also seen faster growth, exacerbating spatial inequalities in national economies. In India, industry and commerce grew largely near and around ports, old port cities emerging as mega metros such as Mumbai, Kolkata and Chennai.

Weaving a Web of Prosperity

In Asia and the Pacific, it is mainly coastal regions that have benefited from the current phase of globalisation by becoming important nodes in the regional production networks. In the context of promoting the Asian Highway and Trans Asian Railway projects, United Nations Economic and Social Commission for Asia and the Pacific (UN-ESCAP) (2010a)examined the potential of intermodal transport and freight modal interchanges and inland ports for alleviating the trade and transit cost disadvantages of land-locked countries and remote hinterlands of littoral states. Extending the developmental impulses to the hinterlands linked to gateways, Asian Highway and Trans Asian Railway are envisaged to serve as catalysts for infrastructural nodes in the region which may take different names, dry ports, ICDs, CFSs, logistics zones or parks, freight villages, distriparks, *et al.* They all will essentially facilitate seamless and cost effective transportation of goods, notwithstanding their varying nomenclature and structure.

ICDs and CFSs, in effect potential hubs of prosperity, stimulate growth of economic activities in their vicinity. Dry ports, as a rule, potentially nurture manufacturing and service clusters, for example, special economic zones and export processing zones. A dry port is generally located close to an existing or potential production or consumption centre. The number of dry ports depend on geography as well as diversity and extent of economic activity. By the mid-1980s about 150 ICDs were established in North America, 130 in Europe and around 400 worldwide (Dobson 1989). In the context of Asian Highway and Trans Asian Railway networks, UN-ESCAP estimated the requirement of an additional 200 dry ports in the Asia-Pacific region by 2015.

ICDs enable and encourage the integration of ports, road and rail freight operations. A network of dry ports as load centres also has the potential to promote traffic on railways, with significant environmental benefits and energy efficiency gains. An optimum use of land transport infrastructure is as a rule derived – long haul between a seaport and an ICD being done by rail and short haul for efficient local distribution, by road. A seaport's access to areas outside its traditional hinterland improves and expands its hinterland. To avoid serious congestion, port operators are increasingly channeling the incoming container flows to satellite terminals or intermodal transfer points in the hinterland.

Gateway Ports to Lean More on Hinterland Nodes

Projections for container port handling worldwide estimate one billion TEU by 2020, doubling the volume of 506 million TEU in 2008 (UNCTAD 2009). With seaport capacities overstretched, more inland container terminals are being developed and expand in capacity. Seaport throughput and efficiency are liable to be jeopardised by bottlenecks in the landside transport system serving the ports. Dry ports have been developed, for example, in the Russian Federation primarily to reduce seaport congestion, whereas in Thailand, priority was to help shift freight from road transport to rail. KTM – Malaysian Railways in conjunction with SRT – State Railway of Thailand arranges rail-based landbridge between Port Klang in Malaysia and Lat Krabang ICD in Thailand. SRT transports substantial quantum of containerised cargo between Lat Krabang ICD and Laem Chabang seaport in Thailand.

7.4   India Ushers in Sustainable Intermodal Growth

Early Start by IR: The First Infant Steps. The development of containerisation in India began with the first ISO container brought to its southwest coast, at Cochin port by an APL vessel on 27 November 1973. It was five years later that the first international container service was introduced on the India–Australia corridor, in September 1978. In India, ICDs and CFSs started to develop as an initiative of Indian Railways (IR), central and state warehousing corporations, shipping lines, logistics and road operators. As early as the 1960s IR realized the immense potential and benefits of high value cargo intermodal door-to-door transport. During this time IR launched a rudimentary multimodal service for "smalls" or less than wagon load traffic through street collection and delivery service. The gradual decline of railroads in most of the industrialised countries was a lesson for IR, to proactively act on fast dwindling general goods traffic. Thus, containerised multi-modal transport in India started as far back as 1966–67 with IR introducing 4.5-tonne and 5-tonne containers of its own design and standards for carrying domestic cargo.

Development of ICDs in India followed a somewhat similar pattern as that in the United Kingdom, following the shipping industry's increasing use of containers for general cargo shipments during the 1960s (see box below).

Box 7.1   Excursus: ICD development in UK

As Ingram (1992) elaborates, the container revolution required customs to recognise ICDs as "ports without water", which according to new customs guidelines had, for example, to be located near trunk roads preferably with access to/from main railway lines. ICDs in the UK were esblished by various consortia, the largest group operating six depots. The Containerbase company, was set up by P&O, five of these depots were in proximity to railway lines linking them to the ports of Tilbury and Southampton by overnight train services. Other ICDs, notably the London International Freight Terminal and Manchester International Freight Terminal were established by British Rail. Some ICDs were developed by companies engaged in warehousing, for example, Greenford ICD (Butlers Warehousing) and Dagenham Storage Co. Ltd. Some others such as Milton ICD in Berkshire were set up by property companies with a view to providing services around which other industries could cluster. Road transport operators also set up some ICDs mainly to bypass the congestion at ports.

In India transportation of ISO containers for international trade from gateway ports to the hinterland was started by IR in 1981–82. In response to the recommendations of different working groups set up by government, IR was required, in consultation with Ministries of Commerce, Finance, and Transport-cum-Shipping, to set up and manage ICDs and develop intermodal transport. Commencing with an improvised ICD at Bangalore within the rail freight handling siding at the station in August 1981, IR created a few other similar ICDs at Coimbatore, Guntur, Anaparti, Amingaon (Guwahati), Ludhiana and Pragati Maidan (New Delhi). The first three provided the linkage for moving containers to and from the ports of Madras (Chennai) and Cochin. Guntur and Anaparti largely for tobacco shipments and the ICD at Guwahati mostly for tea exports. This ICD was linked to the ports of Calcutta (Kolkata) and Haldia. The ICD at Delhi was connected to Bombay (Mumbai) port. India's first inland CFS for consolidation, stuffing and stripping of LCL cargo was commissioned at Central Warehousing Corporation (CWC)'s facility at Patparganj, New Delhi, in 1985.

India put great efforts to drive containerisation in its international trade in the 1980s–1990s. Until lately India's container traffic amounted to just about one per cent of global container volumes, albeit, recently, it has shown a slight increase and is expected to grow substantially.

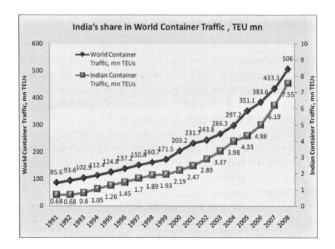

Figure 7.1   India's share in world container traffic, million TEU
Source: Based on data from Containerisation International Yearbooks

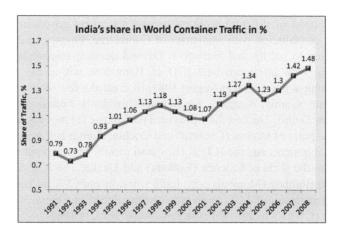

Figure 7.2   India's % share in world container traffic
Source: Based on data from Containerisation International Yearbooks

End-to-end Multimodal Transport

The share of container traffic in IR's total freight traffic in 1988–89 was miniscule: of a total IR's freight loading of 302 million tonnes, containerised traffic accounted for less than 0.5 million tonnes. International trade required more transit-time sensitivity and care for "small-volume customers" care, which required IR to change the focus from large volume bulk commodities transport. Trade facilitation may well be cited as the very *raison d'être* of containerisation of India's export–import cargo. Particularly, for the ease of handling and safe and speedy transit. Further, overseas importers insisted on receiving their shipments in containers, which facilitated of door-to-door transport backed by a through, unified liability regime, a composite contract with the forwarder involving multimodal transport end-to-end. This drove the establishment of inland terminals close to the sources of the cargo. A customs clearance facility was a further critical element.

8.081 million TEU were handled at all ports in 2009–10, including over 5.4 million at Jawaharlal Nehru Port (JNP). Combined JNP and Chennai ports handled 77 per cent of all TEUs.

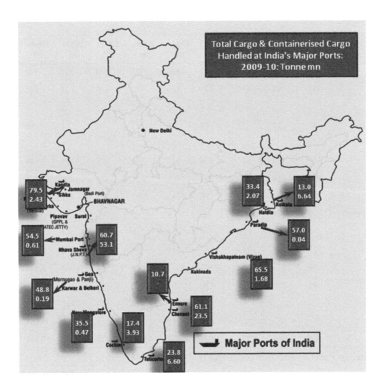

Figure 7.3   Important container handling ports in India, 2009–10
Source: Annual Statistics from Indian Ports Association

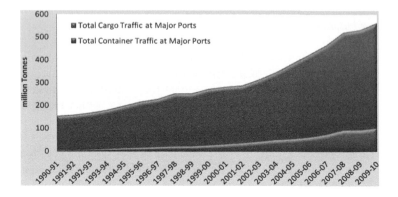

Figure 7.4   Share of container traffic in total traffic at major ports (million tonnes)
Source: Data from Indian Ports Association

Table 7.1   Containers handled at Indian ports, 2000–01 to 2009–10 (thousand TEU)

	2000–01	2001–02	2002–03	2003–04	2004–05	2005–06	2006–07	2007–08	2008–09	2009–10
Kolkata	138	98	106	123	159	203	240	297	302	378
Haldia	51	93	117	137	128	110	110	128	127	124
Paradip	-	-	2	4	2	3	2	4	2	4
Visakhapatnam	20	22	22	20	45	47	50	71	88	97
Ennore	-	-	-	-	-	-	-	-	-	-
Chennai	352	344	425	539	617	734	798	1128	1,144	1,216
Tuticorin	157	214	213	254	307	321	377	450	439	440
Cochin	143	152	166	170	185	203	227	254	261	290
New Mangalore	2	4	6	7	9	10	17	21	29	32
Mormugao	4	6	9	10	10	10	12	14	14	13
Mumbai	321	254	213	197	219	156	138	118	92	58
JNPT	1189	1573	1930	2269	2371	2668	3,298	4,060	3,953	4,092
Kandla	91	126	157	170	181	148	177	165	137	147
Mundra				49	211	299	529	712	795	909
Pipavav				25	69	86	153	155	168	281
Total	**2,468**	**2,886**	**3,373**	**3,900**	**4,517**	**4,998**	**6,128**	**7,577**	**7,551**	**8,081**

Source: Indian Ports Association/other ports

Table 7.2   Total cargo and "others" at India's major ports, 2009–10

Ports	Total Cargo (million tonnes)	Others	General Cargo Containerised (million tonnes)	Total	Share of general cargo *vis-à-vis* Total Cargo (%)	Share of containerised cargo in general cargo (%)
All major ports	561.09	95.01	101.24	196.25	35.0	51.6
Kolkata	13.04	5.38	6.64	12.02	92.2	31.6
Haldia	33.38	6.49	2.07	8.56	25.6	24.1
Visakhapatnam	65.50	11.18	1.68	12.86	19.6	13.1
Ennore	10.70	0.09	-	0.09	0.8	-
Chennai	61.06	12.56	23.48	36.04	59.0	65.1
Tuticorin	23.79	8.95	6.60	15.55	65.4	42.4
Cochin	17.43	1.05	3.93	4.98	28.6	78.9
New Mangalore	35.53	3.01	0.47	3.48	9.8	13.5
Mormugao	48.85	2.25	0.19	2.44	5.0	7.8
Mumbai	54.54	15.25	0.61	15.86	29.1	3.8
JNPT	60.76	2.75	53.10	55.85	91.9	95.1
Kandla	79.50	20.27	2.43	22.70	28.6	10.7
Paradip	57.01	5.77	0.04	5.81	10.2	0.7

Source: Indian Ports Association

The western and northwestern hinterland account for a predominant share in export–import containerised cargo, 68 per cent of India's container traffic handled in the country; the southern region has a share of 25 per cent; the eastern region accounts for just about 7 per cent accordingly.

Institutional Framework

Four important prerequisites were identified that had to be met to establish and develop ICDs/CFSs: (i) appropriate location determined by an overall assessment of the business potential; (ii) coordination by and support of central government, involving also state and local government, wherever required; (iii) legal and regulatory framework; and (iv) integrated transport infrastructure serving the nodes. For a coordinated and concerted thrust to containerisation of India's export–import trade, an Inter-Ministerial Committee (IMC) with representation from the Departments of Commerce, Revenue (Customs), Shipping and Ports and Railways was founded. At that time thirty ICDs and CFSs, of which 28 were in the public sector, had already started operations. Since the implementation all applications for setting up ICDs and CFSs are analysed and approved by IMC. A feasibility study precedes the proposal for an ICD/CFS which has to constitute the economic viability of the facility and its attractiveness to users and a minimum critical mass.

ICDs/CFSs are expected to provide for all trade-related facilities; single-window for mandatory clearances, payments, and incentives certification, and presence of customs, banks, shipping lines and agents, NVOCCs (non-vessel owning common carriers), CHAs (customs house agents), and transport operators. Following an indicative norm expected minimum throughput for a rail-linked ICD is 6,000 TEU and 1,000 TEU for a satellite feeder CFS. Further, the IMC expects assurance of availability of land for future expansion in accordance to the assessed import and export traffic flows. Following the facility's infrastructure development as approved by IMC, a notification is issued under the Customs Act declaring the facility as a customs area and the operator as a custodian.

Assessing Potential and Flows

To determine the operational and economic viability of an inland terminal, its traffic potential and development prospects are taken into consideration. According to a survey from Rail India Technical and Economic Services Ltd (RITES) (2002), a leading consultant, it was first necessary to assess both existing and the future container cargo flows and to establish the related inland origins (export) and inland destinations (import) of such traffic to properly define the demand for container facilities (ICDs/CFSs) in the next 15 years. The number and size of dry ports would depend on geography as well as diversity and the extent of economic activity and the industrial production and commercial transactions in the area served by the facility. RITES (2002) observed that the demand elasticity of container traffic with regard to GDP for the 5-year period (1995–96 to 2000–2001) was 1.98 (GDP rose at around 6 per cent CAGR, container traffic grew at over 11 per cent). Based on the premise of a minimum of 40,000 tonnes of containerised export/import cargo (equiv.to 6,000 TEU per annum) it identified 197 locations for the establishment of ICDs, including the existing ones.

Legal, Liability and Facilitation Aspects

In order to keep important features relevant to legal, liability, financial issues and electronic data interchange in focus the Government enacted the Multimodal Transport of Goods Act (MTGA), which served as the basis for legal support for international banking and liability aspects. Until the enactment of this Act, the Foreign Exchange Dealers Association of India (FEDAI) evolved its own rules, laying down responsibilities and liabilities of the combined transport operators. The FEDAI did not confer negotiability title to the goods. Further, such documents were required to be exchanged for regular on-board Ocean Bill of Lading (OBL) at the port, unless the Letter of Credit (LC) specifically permitted the production of the Combined Transport Document (CTD) evolved by FEDAI in relation to the Bill of Lading. Insurance by way of liability cover for MTOs could be extended by the TT (Through Transport) Club. The multimodal transport document serves as an instrument to enforce the provisions of the Act by assigning liabilities and responsibilities to MTO, consignor, consignee, insurer and banker.

A few other measures towards trade facilitation by customs authorities entailed a simplified procedure for amendment of IGM (inland general manifest), simplified customs procedures for transhipment between the gateway port and the dry port, allowing LCL containers to be moved from one CFS to another for final consolidation/stuffing, as well as a facility for exchanging customs messages with ports, airports, ICDs/CFSs, intermodal operators, banks and the DGFT (Directorate General of Foreign Trade), and a facility for payment of Customs duty via banks and e-banking.

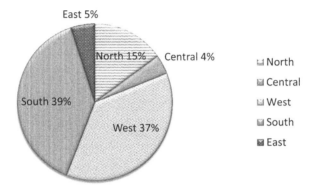

Figure 7.5   Approvals by IMC for setting up of ICD/CFSs as of 31 March 2010
Source: Ministry of Commerce and Industry, GoI

7.5   CONCOR's Comprehensive Container Culture

The Birth of CONCOR: A Seminal Step

The emerging importance and potential of containerisation made it essential to install a specialised public sector entity with focus on the creation of infrastructure for multimodal containerised transport. This also required the essential institutional support and involvement of relevant stakeholders including customs, railways, ports, central bank, shipping lines and industry associations. Accordingly, the government of India set up a separate government-owned corporate body for the facilitation and promotion of multimodal transport. Container Corporation of India Ltd (CONCOR) was incorporated in March 1988. CONCOR had an important task to manage the change in India's logistics architecture, to spearhead a container culture in the country, to build and operate infrastructure linkages for rapid and accelerated inland penetration of containerised international trade traffic, and to promote containerisation of intra-country domestic general cargo.

CONCOR commenced by establishing a string of ICDs and CFSs for cargo collection and delivery and its consolidation and disaggregation. CONCOR's pioneering role may well serve as a tangible case study for the development of containerisation, establishment of inland container handling infrastructure, intermodal logistics and promoting institutional mechanisms for public–private partnership favourable to sustainable growth of multimodal transport.

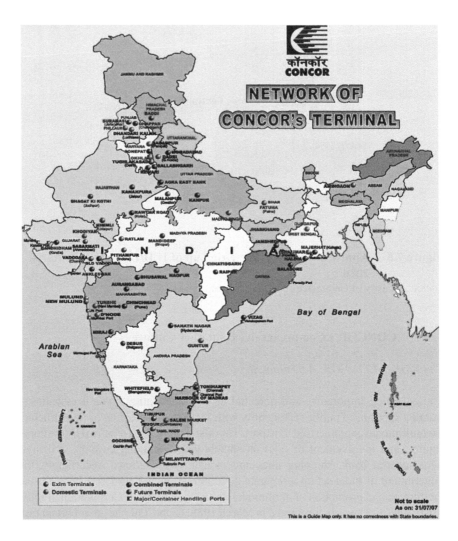

Figure 7.6   CONCOR dry port network
Source: CONCOR

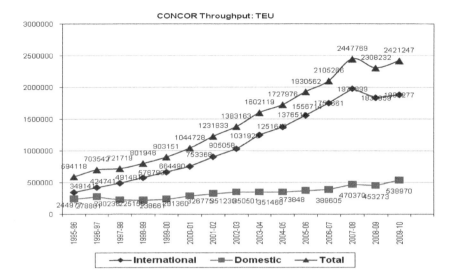

Figure 7.7   CONCOR throughput (TEU)
Source: Data from CONCOR

Generating Container Culture

CONCOR's growth strategy, to improvise and build in stages; helped to conserve resources, cut costs of operations, and built speedily. While it steadily engendered the multimodal milieu in the industry, the concomitant institutional framework designed by government helped expand and sustain its growth. Due attention was paid to the development of the "software" part of multimodalism (trade facilitation, systems, procedures, documentation, compensation claims liability, service concepts, etc.)

A Financial Success

Steadily moving on a high growth path, in the range of 13 per cent to 15 per cent annually, CONCOR contributed to the shareholders' wealth. It started with authorised capital of Rs. 100 crore (1 billion) and paid up capital of Rs. 65 crore (650 million); 37 per cent of its equity share capital has subsequently been disinvested; CONCOR has thus evolved as a mixed public–private company, with significant private equity. Its current market capitalisation exceeds Rs. 17,000 crore (almost US$ 4 billion).

Although 63 per cent of its ownership still vests in the Government of India, CONCOR remains an autonomous commercial enterprise with its own Board of Directors wholly responsible for its commercial activities and financial results. It receives neither subsidy, nor any exclusive consideration from government.

Besides the sizable annual dividend (Rs. 115 crore in 2009–10) for IR's equity investment in CONCOR, the latter pays IR lease charges for lands utilised for constructing infrastructure by way of income share per TEU handled at its ICDs/CFSs, in addition to full freight for its container trains operatd by IR. These are in no way dissimilar from the rates levied on on other container train services. In 2009-2010 it paid Rs. 3,886 crore (38.86 billion) as freight to IR for its container trains operated on the IR system.

Customer Focus: Suite of Services

Most ICDs provide single window facilities, cargo aggregation and storage, palletisation, container stuffing and de-vanning, customs clearance, truck and trailer parking, banking and office space for customers' agents/employees regularly working at the terminals. Right from its inception, CONCOR visualised the CFSs to be trade growth centres and included the following facilities, some of them in partnership with the private sector:

- Container repair and cleaning facilities
- Cargo palletisation, strapping, lashing/choking, etc.
- Fumigation of cargo/containers
- Door delivery/pick-up of containerised cargo
- Container/cargo survey
- Flexible payment arrangements
- Provision of reefer facilities.

As if to follow the words of the celebrated astrophysicist Stephen W. Hawking: "Time is not completely separate from and independent of space, but is combined with it to form an object called space–time, in relation to motion", CONCOR devised its business strategy to be in harmony in the trinity of time, space and motion. CONCOR pursued the essence of intermodalism with the primacy of rail transit and road transport as supplementary services (with involvement of the private sector) to provide door-to-door linkages. A simplified tariff system, the ingenious Inland Way Bill (IWB), was a composite charge for most of the common services including terminal handling charges at the ICD and at the gateway port together with rail haulage cost for containers between the seaport and the ICD. The composite freight charge on the basis of per TEU km formed a virtual FAK (freight all kind) irrespective of the commodity in the box.

CONCOR pursued a commitment towards a slim structure, multi-skilled personnel, and a generally austere regime. Lofty, lumpy investments were eschewed; instead, sights were generally set on facilities which would come up speedily and on improving infrastructure, to begin with. That enabled several schemes and projects, awaited for years, to be put on the ground expeditiously.

Direct stuffing of goods in containers straight from the road vehicle without first being unloaded in a warehouse and inspected by Customs helped to save time as well as cost. Likewise, the Central Excise staff posted for other regulatory functions in the shippers' premises supervised stuffing of containers and sealed them on behalf of Customs. This proved to be an important element in progressively increasing the factory stuffing/destuffing of export/import containers, thereby helping door-to-door multimodal transport. Mini-bridging of import containers from port to port by rail helped ships avoid time and expense involved in calling at multiple ports within India. Bonded warehousing for storage of import cargo enabled fast release of import containers which, in turn, improved productivity of terminals. The addition of domestic containerised rail transport acquired a prominent place in its business.

The concept of the Port Side Container Terminal (PSCT), akin to a near-dock facility, has been an ingenious concept. A PSCT was commissioned in March 1991 at Tondiarpet, within 6km from the port of Chennai; Wadi Bander came up the following month across the way from Mumbai port, and today Dronagiri Node is in operating near Jawaharlal Nehru port (JNP). In a similar way, a few other PSCTs added a new dimension to container handling. They complemented the facilities offered by premier container handling seaports, particularly Chennai Port on the east coast and JNP and Mumbai ports on the west coast. The PSCTs served as intermodal hubs concomitant to quayside operations, facilitating quick dispersal of import containers from ports, and efficient aggregation of export containers for timely loading on to vessels, relieving congestion and reducing strain on the port infrastructure, in fact, helping optimise port capacity utilisation. The concept further facilitated the rail land-bridge development to link specialised ports between the east coast and the west coast.

The Bicycle Wheel

CONCOR has pursued "hub and spoke" operations for road or short haul rail shuttle services within defined catchment areas. Some hubs like the ICD at Tughlakabad (TKD) in New Delhi were fed by several satellite locations such as Panipat (c. 100 km) or even Gwalior (c. 300 km), until traffic steadily built up and justified running a scheduled container train service from the satellite facility itself. This was also done in the case of Ludhiana and Moradabad, which started out as remote locations linked to the hub terminal at TKD. As traffic volumes increased, these satellite terminals were expanded. They soon started functioning as stand-alone terminals, operating through trains from and to gateway ports such as JNP. These satellite CFSs were served by road transport, following a road-chassis system; for instance, at the CFSs at Moradabad and Panipat, the former an important node for handicrafts, and the latter for textile products. Empty containers from ICD at TKD would be taken on road trailers, export cargo customs-cleared and stacked in the respective facility in advance, to be directly stuffed into the container, while still on chassis, and the sealed container moved back to the rail container complex at TKD, all in a matter of few hours, for onward dispatch to gateway port by an express block train.

With road linkages between the satellite CFSs owned and managed by CONCOR, the aim was to provide a comprehensive network to cover most of the major trade centres in the country. Most of the CFSs and all ICDs owned by CONCOR in combination with the road/rail linkages enables the company to offer a complete service package unlike that availalbe in UK and Germany, for example, where Freightliner and Transfracht, respectively, operate intermodal services between ports and inland centres owned and operated by others.

Involving Airlines

Generally, ICDs and CFSs have traditionally been linked only to maritime services and ports. CONCOR also included air cargo in its ambit. It joined hands with some airlines and started handling air cargo at its ICDs, customs-cleared ULDs being carted between CONCOR ICDs and gateway airports in customs-bonded trucks. CONCOR has joined hands with Hindustan Aeronautics Ltd at Bangalore through a revenue sharing model, to develop facilities for airfreight of export cargo. It has also started coastal shipping in association with Seaways Ltd.

Synergising Public and Private Strengths

As the debate constantly rages for the desired option of private sector participation in management and operation *vis-à-vis* government or public sector entities, at times the public–private partnership (PPP) format even becomes a fetish. CONCOR joined hands with a number of private sector as well as other public sector companies, blending their mutual synergies and strengths with the objective of deriving optimal advantage, cost reduction, and efficiency enhancement. For example, the CFS at Mulund (in Mumbai) set up by CONCOR as a joint venture with a local labour cooperative society was an innovative approach and helped operate the facility efficiently as a near-dock terminal for Mumbai port and Jawaharlal Nehru port. CONCOR believed that a PPP constitutes a sustained collaborative effort between the public sector and private enterprise in which each partner shares in the design of a project, contributes a portion of the financial, managerial and technical resources needed to design and execute that project, and partially shoulders the risks and obtains the benefits that the project creates.

In the case of setting up of satellite CFSs, participation of agencies like the state and central warehousing corporations and those in the private sector was sought at locations where appropriate warehousing infrastructure was available. In the interim-phase development, cargo destuffing of import containers, customs inspection and delivery was arranged at some of the customs-bonded warehouses of these agencies as at Ludhiana, Jalandhar, Amritsar, Ahmedabad, Pune and Hyderabad. Private entrepreneurs were invited and encouraged to join hands in different activities on mutually acceptable terms, for instance, in providing capital and trained manpower for handling equipment (cranes, trucks, forklifts, etc.), at its ICDs and CFSs, maintenance facilities, and terminal operations. Besides, CONCOR proposed to let the private operators handle containers and road cargo between satellite CFSs, the rail-fed ICDs and shippers' premises on contract/under franchise. Nevertheless, at selected facilities such as its flagship ICD at Tughlakabad, CONCOR owns, operates and maintains an array of its own equipment.

At Dadri, near Greater Noida, in the vicinity of Delhi, CONCOR has joint ventures (JV) to develop CFSs with four shipping lines: Maersk, APL (later switched to Albatross), CMA CGM, and Transworld – with 49 per cent equity participation. It has also entered into a JV with Maersk, with 26 per cent equity contribution for the development and operation of the third container terminal at JNP. The 1.3 million TEU terminal became operational in March 2006. Another JV was entered into with Dubai Port World with 15 per cent equity contribution to develop a container transhipment port at Vallarpadam in Kochi. Yet another joint venture has been forged between CONCOR and Gateway Rail Freight Pvt. Ltd, a subsidiary of Gateway Distriparks, to construct and operate a rail-linked container terminal at Garhi Harsaru near Gurgaon in proximity of Delhi.

The country faces a serious challenge from the need of substantial capacity enhancement at gateway ports as well as dry ports along with the multimodal transport networks. With resources being scarce in the state domain, a clear and conscious emphasis is on actively promoting public–private partnerships. The Ministry of Railways has provided support for private container train operators on IR network, involving 16 players (including CONCOR). This, in effect, is expected to drive an expansionary phase for multimodal transport in the country. IR has created Dedicated Freight Corridor Corporation of India Ltd (DFCCIL), and planned to develop freight terminals, and logistics parks on PPP basis.

The list of ICDs and CFSs approved by the Inter-Ministerial Committee amply demonstrates that contrary to a perception in some circles, there has been a strong participation of the private sector in the development of ICDs and CFSs in India. As many as 155 of all 246 ICDs and CFSs approved by IMC (up to 31 March 2010) are in the hand of the private sector. The size of the Indian container logistics market comprising ICDs and CFSs is estimated to have a value of Rs. 6,500 crore or 65 billion (Container Logistics, CFSs and ICDs India, April 2009).

Potential of Intermodal Growth

There is immense scope and potential for containerisation growth in the country. Textiles contribute about 18 per cent of India's total exports; this share is expected to increase to 25 per cent and more. Engineering industry contributes more than 11 per cent of total exports; this share is expected to rise to 15–18 per cent by 2012. India's auto component exports rose by more than 33 per cent in 2005–06 and are likely to maintain the tempo in the future. A.T. Kearney (2008), among many other expert bodies, has estimated that India's retail industry will grow to over $ 200 billion, recording a 30 per cent CAGR over the next few years. Along with significant demographic transformation, increasing disposable incomes impact people's consumer behaviour and lifestyle in emerging economies. The expected increase in value-added exports, the level of containerisation will substantially improve.

Asia Moves Ahead

Of the three major trade routes Asia–Europe, trans-Pacific and intra-Asian routes, the intra-Asian trade is projected to show the strongest growth rate until the year 2015. UNESCAP (2010) expects the total volume of international container handling in the Asia-Pacific region ports to increase from 142.7 million TEUs in 2002 to 427.0 million TEUs in 2015, an annual average growth rate of 8.8 per cent. The trade triangle comprising India–China, India–ASEAN and ASEAN–China is projected to account for about 80 per cent of Asia's container trade by 2025. A clear indication of a new geography of trade.

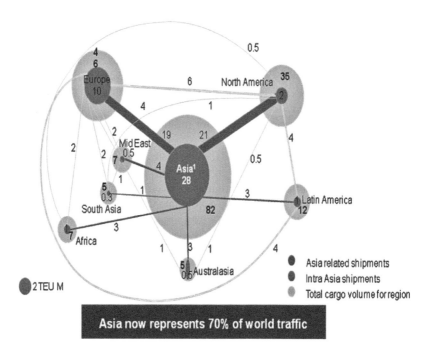

Figure 7.8   Asia now the global centre for cargo
Source: Drewry, BCG

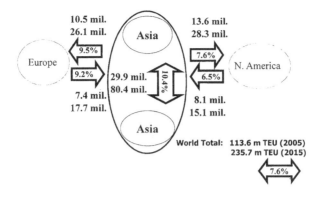

Figure 7.9   Projected container trade, 2005–2015
Source: UNESCAP

7.6   Inadequacies to be Addressed

Adequate Landside Connectivity Critical

Seaports rely heavily on inland ports to preserve their attractiveness. Their performance is closely entwined with the development and performance of associated inland networks that provide access to cargo terminals in the hinterland. Distances from the prominent container ports on the west coast linking important ICDs in the hinterland, especially in the north and north western parts of the country, are indeed long.

Table 7.3   CONCOR's share in port traffic (TEU)

Port	2009–10			2008–09			2007–08		
	Port throughput	CONCOR throughput	% Share	Port	CONCOR	% Share	Port	CONCOR	% Share
Chennai	1,225,000	103,955	8.49	1,144,000	134,771	11.78	1,128,000	151,075	13.39
JNP	4,061,000	1,001,553	24.66	3,953,000	946,649	23.95	4,060,000	1,029,981	25.37
Mundra	909,238	94,496	10.39	795,276	70,341	8.84	711,550	66,007	9.28
Pipavav	280,867	98,845	35.19	168,415	55,564	32.99	155,416	73,780	47.47
	6,476,105	1,298,849	20.06	6,060,691	1,207,325	19.92	6,054,966	1,320,843	21.81

Source: CONCOR

A considerable volume of containerized cargo received at ports gets customs-cleared at CFSs in their vicinity and is thereafter transported break-bulk by road, however it is not uncommon that even full containers are carried by road owing to paucity of rail capacity. More than 28 per cent of the containerised cargo moving by road is carried over distances in excess of 300 km.

Table 7.4   Rail and road distances (km). Rail distances are based on Railway Time Tables, road distances on Indian Distance Guide (TTK Publication)

From	To	Dhandari Kalan (Ludhiana)	Morad-abad	Agra Cantt	Dadri	Tughlak-abad	Sabar-mati	Gwalior (Malanpur)	Vadodara
Pipavav	Road	1,537	1,369	1,029	1,266	1,190	337	1,205	341
	Rail	1,593	1,469	1,230	1,332	1,319	431	1,373	531
Mundra	Road	1,461	1,397	1,130	1,241	1,312	425	1,280	538
	Rail	1,528	1,388	1,165	1,267	1,239	356	1,283	456
JNP	Road	1,713	1,566	1,516	1,442	1,387	561	1,251	442
	Rail	1,741	1,588	1,293	1,464	1,413	580	1,436	437

Table 7.5   Container traffic at "major" ports: growth trends (thousand TEU)

	All India		West Coast		Mumbai Region		JNP		JNP Share in India overall (%)
	Traffic	Growth Rate (%)	Traffic	Growth Rate (%)	Traffic	Growth Rate (%)	Traffic	Growth Rate (%)	
1995–96	1,449	15.27	1,182	26.68	857	17.23	339	38.93	23.4
1996–97	1,698	17.18	1,198	1.35	1,006	17.39	423	24.78	24.9
1997–98	1,892	11.43	1,314	9.68	1,105	9.84	504	19.15	26.6
1998–99	1,926	1.80	1,369	4.19	1,178	6.61	669	32.74	34.7
1999–2000	2,223	15.42	1,565	14.32	1,320	12.05	890	33.03	40.0
2000–2001	2,532	13.90	1,743	11.37	1,512	14.55	1,190	33.71	47.0
2001–02	2,899	14.49	2,120	21.63	1,828	20.90	1,574	32.27	54.3
2002–03	3,344	15.35	2,461	16.08	2,142	17.18	1,929	22.55	57.7
2003–04	3,906	16.81	2,822	14.67	2,465	15.08	2,269	17.63	58.1
2004–05	4,233	8.37	2,975	5.42	2,590	5.07	2,371	4.50	56.0
2005–06	4,613	8.98	3,195	7.39	2,824	9.03	2,668	12.53	57.8
2006–07	5,541	20.12	3,871	21.16	3,436	21.67	3,298	23.61	59.5
2007–08	6,596	19.04	4,625	19.48	4,176	21.54	4,059	23.07	61.5
2008–09	6,588	-0.12	4,045	-3.28	4,045	-3.14	3,953	-2.61	60.0
2009–10	6,891	4.60	4,632	3.55	4,150	2.60	4,092	3.52	59.4

Source: Jawaharlal Nehru Port

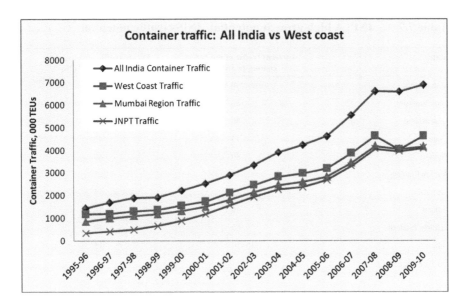

Figure 7.10   Container traffic: all India vs West Coast
Source: Statistics from Indian Ports Association, CONCOR, JNP

With a share of 60 per cent of the entire container traffic handled at India's "major" ports, JNP commands the pride of place. Although its potential is reckoned to be enormous as determined by a recent KPMG analysis, its optimal annual capacity is pegged at 8–10 million TEU mainly on account of inadequate draught and other constraints on the water front restricting its expansion further.

Table 7.6   Distant ICDs: JNP's prominent users

JNP: Rail share of container traffic: 2007–08, principal ICD-wise		
	Total rail-carried (TEU)	Share for ICD in total JNP's rail-borne traffic (%)
Tughlakabad	356,695	35
Dadri	128,972	13
Dhandari Kalan	110,274	11
Sabarmati	66,420	7
Nagpur	51,663	5
Loni	50,140	5
Total of six ICDs	764,164	76
Others	245,363	24

Source: JNP data analysed

Table 7.7   JNP: A high growth potential. JNP's traffic potential

State	Current traffic of the region coming to JNP (%)	Future traffic of the region coming to JNP (%)
Maharashtra	90	80
Uttar Pradesh	80	40
Uttarakhand	80	40
Delhi	75	40
Punjab	80	40
Andhra Pradesh	10	5
Karnataka	37	5
Gujarat	60	10
Madhya Pradesh	60	40
Others	15–20	15

Source: KPMG

Table 7.8   Estimates of container traffic at JNP (thousand TEU)

	JNPCT	NSICT	GTICT	Terminal-IV (Proposed)	Total
2003–04 (Actuals)	1,038	1,231	-	-	2,269
2006–07	1,308	1,231	400	-	2,939
2011–12	1,801	1,231	1,065	750	4,847
2016–17	2,000	1,231	1,300	2,420	6,951
2021–22	2,000	1,231	1,300	3,500	8,031

Source: RITES (Rail India Technical and Economic Services)

One likely alternative is that ports like Mundra and Pipavav to increasingly handle containers originating from and terminating at ICDs in the northern/north western region and JNP largely handling the traffic to, and from, central and south central regions besides the metropolitan Mumbai region itself. Mundra and Pipavav ports, being linked by non-electrified rail lines are able to operate double stack container trains, thereby enjoying economy of scales in addition to increase in rail transport capacity.

Imperative of Intermodal Capacity Development

There is a clear imperative of developing adequate railway infrastructure in the context of EXIM container trade to provide connectivity between hinterland demand-production centres and key gateway ports as an efficient means of clearing landside port areas and transporting containers over long distances on land. Hinterland potential for export–import container traffic handled at ports is estimated to be at least 70 per cent; actual movement of full containers from and to hinterland locations is currently less than 35 per cent. Rail-borne container movement between ICDs and gateway ports currently is less than 25 per cent; an optimal ratio would be about 50 per cent. Inland movement of containers by rail over long distances inherently offers economies of scale and associated benefits of potentially lower transportation costs. (McKinsey and Company, 2010). Inadequate capacity on IR's western corridor seriously hampers JNP's container traffic to and from hinterlands.

The share of road transport amounts to around 66 per cent of the total movement by all modes. Over 60 per cent of containers movement by road is for distanaces of up to 150 km. Another 11 per cent are moved over distances between 151 and 300 km. The share of container movements over distances greater than 300 km is around 28 per cent; these can be as long as 884 km average.

7.7   Immense New Potential on the Horizon

Vast New Opportunities Unfold

Recognising the availability of quality infrastructure as the prime requirement to realise full economic growth potential ging, India's dedicated multi-modal high axle load freight corridors along the country's golden quadrilateral together with its diagonals, linking the metropolises of Delhi, Mumbai, Kolkata and Chennai, constitute an ambitious project. Phase I of this planned development involves the construction of Delhi–Mumbai and Delhi–Kolkata rail freight corridors, both totaling about 2,800 km. The western corridor is being planned to operate double stack container trains with electric traction.

The dedicated freight corridor (DFC) between Delhi and Mumbai, connecting Vadodara–Ahmedabad–Palanpur–Marwar Jn–Phulera–Rewari, will cover a length of 1,483 km with end terminals at Tughlakabad and Dadri (both already serving as India's largest ICDs) in the National Capital Region of Delhi at one end and JNP at Mumbai at the other. Traffic for the Mumbai–Delhi DFC is projected at 38 million tonnes in 2013, rising to 106 million tonnes in 2023 and 157 million tonnes by 2033. To a considerable extent, traffic growth on this rail corridor will be driven by rising international container traffic through the ports on the west coast, containers comprising nearly 80 per cent of overall throughput. Of a total 15.5 million TEUs to be handled at India's ports in 2021–22, rail mode is expected to carry 6.1 million TEUs.

Table 7.9 Projected traffic growth on the western corridor, trains per day with current axle load and single-stack container operation

	2005–06		2011–12		2016–17		2012–22	
	Up	Down	Up	Down	Up	Down	Up	Down
Containers	14.8	12.3	56.4	50.6	82.5	78.5	108.9	102.5
Coal	-	1.4	-	3.4	-	4.7	-	6.2
Fertiliser	1.1	4.1	1.3	4.6	1.4	5.1	1.5	5.6
Salt	0.5	2.2	0.6	2.5	0.6	2.7	0.7	3.0
Cement	0.1	0.9	0.2	1.2	0.2	1.5	0.3	1.9
Iron and Steel	0	0.4	0	0.5	0	0.7	0	0.8
POL	1.5	4.4	1.7	4.9	1.9	5.4	2.1	6.0
Foodgrains	4.1	0.2	4.6	0.2	5.1	0.2	5.6	0.2
Misc.	0.6	1.5	0.8	1.7	0.9	2.0	1.1	2.3
Empties	6.8	2.6	8.1	3.4	9.4	4.0	10.9	4.6
Total	29.7	30.1	73.7	72.9	102.1	104.7	131.1	133.2

Source: RITES

This DFC along the western corridor will traverse six of India's states, viz., the NCR, Uttar Pradesh, Haryana, Rajasthan, Gujarat and Maharashtra, which together constitute 39 per cent of India's total population, contribute 60 per cent of its exports, 54 per cent of overall industrial output, and 43 per cent of national income. Passing through the region that already has well-developed industrial base, the development of the western DFC is thus envisaged to serve, in turn, as a catalyst for the US$90 billion Delhi Mumbai industrial corridor (DMIC).

Investment Regions (IRs) and Industrial Areas (IAs) are proposed to be developed as self-sustained investment nodes duly equipped with high speed multi-modal transportation and logistics systems. The concept of IRs/IAs is strengthened and justified by the cluster approach of Michael Porter (XXX): "A cluster is a geographically proximate group of companies and associated institutions in a particular field, linked by commonalities and complementarities."

The DMIC project lends a special importance to the whole panoply of multimodal linkages: development of logistic hubs, container freight stations and bonded warehousing zones – all to ensure end-to-end solutions for optimal freight distribution through multi-modal transport, provision of requisite cargo/ container warehousing and handling facilities along with storage, parking, repairing, servicing and transhipment wherewithal.

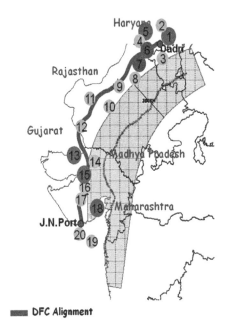

S.No	State	Location	Category
1	Uttar Pradesh	Dadri-Noida-Ghaziabad	IR
2	Uttar Pradesh	Meerut-Muzaffarnagar	IA
3	Haryana	Faridabad-Palwal	IA
4	Haryana	Rewari- Hissar	IA
5	Haryana	Kundli-Sonepat	IR
6	Haryana	Manesar-Bawal	IR
7	Rajasthan	Kushkhera-Bhiwadi-Neemrana	IR
8	Rajasthan	Jaipur-Dausa	IA
9	Rajasthan	Ajmer-Kishangarh	IA
10	Rajasthan	Rajsamand-Bhilwara	IA
11	Rajasthan	Pali-Marwar	IA
12	Gujarat	Palanpur-Sidhpur-Mahesana	IA
13	Gujarat	Ahmedabad-Dholera	IR
14	Gujarat	Vadodara-Ankleshwar	IA
15	Gujarat	Bharuch-Dahej	IR '
16	Gujarat	Surat-Navsari	IA
17	Gujarat	Valsad-Umbergaon	IA
18	Mahrashtra	Nashik-Sinnar	IR
19	Mahrashtra	Pune-Khed	IA
20	Mahrashtra	Alewadi/ Dighi Port	IA

Figure 7.11   Proposed investment regions and industrial areas
Source: Infrastructure Development Corporation Limited

NCR: An Economic Hub of High Promise

The National Capital Region has evolved into a major manufacturing and trading hub as well as a major consumption centre. IL&FS (2009) surveys (*Multimodal Logistics Park at Rewari*) reveal that DMIC–related developments in the NCR and Manesar–Bawal clusters will generate the expected large container volumes estimated to be at a level of at least 4 million TEUs by 2020. The cumulative capacity at all existing dry port facilities in the region being at best 2.5 million TEUs, there is a need for a large ICD to come up at a place like Rewari where the proposed dedicated rail freight corridor and the multimodal logistics park will be able to leverage each other. The rail-linked ICD at Rewari, about 85 km from New Delhi, is envisaged to have an automobile park-cum-loading facility for rail-borne transportation of automobile traffic for export as well as domestic markets, a freight consolidation centre, an industrial centre, a trade centre, ancillary services such as customs bonded warehousing, cold storage, and value-addition services.

A Panoply of Logistics Infrastructure

The Indian logistics industry is expected to grow annually at a rate of 15–20 per cent, reaching revenues of approximately $385 billion by 2015 (Cushman Wakefield 2008). High concentration of retail, special economic zones, and the emerging manufacturing hubs/clusters are triggering unprecedented levels of logistics infrastructure to be installed across the country. Some 110 logistics parks with a total of about 3,500 acre of land and estimated cost of US$ 1 billion as well as 45 million sq ft of warehousing space with an investment of US$ 500 million are expected to be operational by 2012. Most of these developments will be concentrated in 14 locations, including several tier-two and tier-three cities and peripheral locations such as Bhiwadi on the outskirts of Mumbai, Panvel in proximity of JNP, Haldia, Falta, Dankuni, Kharagpur and Durgapur in West Bengal, besides those near Hyderabad, Chennai, Visakhapatnam, Nagpur, Gurgaon and Kochi.

7.8   Imperatives of Mid-course Correction

Mushrooming CFSs around Ports

Almost half of the country's inland container terminals are located in port towns and around the gateway ports, for example, 34 in the Mumbai area, most of them in proximity of JNP, 29 around Chennai port, 18 close to Kolkata/Haldia ports, 15 in Tuticorin, 12 in Mundra. Given the lucrative business many more applications for CFS than desirable have made.

The mushrooming of CFSs in port towns, as in the case of JNP, is driven by the delays in transporting import containers by rail to ICDs in the hinterlands. Further, some clearing/forwarding agents with traditional commercial links to trade in the port town itself remain committed to CFSs in the vicinity. CFS operators are reported to be paying a premium to the shipping lines for getting the import containers nominated to their CFSs. CFSs servicing JNP are estimated to have handled some 2.10 million of import containers in 2009–10.

Skewed Dry Port Distribution: Need for a Buoyant Gateway on East Coast

The skewed distribution of dry ports, with the eastern region having only a one per cent share, is an anachronism as well. India's avowed "Look East" policy and a legitimate expectation of a surge in India's intra-Asian trade, point to the desirability of concerted initiatives for holistic development of a large container port on the east coast with strong landside connectivity.

Pragmatic Rail Transport Pricing

The freight haulage charge policy of IR is only little competitive with road prices, particularly in respect of light weight cargo, making container penetration into the deep hinterland uncompetitive. Light and bulky cargo laden forty foot containers do seldom get on rail; they are more often dealt with at port-side CFS facilities.

Large Common user ICDs

For container train operations, the hub-spoke model appears optimal. There is need for about 10–12 large hub terminals on a pan-India network, supported by 30–40 spoke terminals. The size of the ICD will need scaling up so as to provide choices that are currently available in a port, and not at small sized ICDs in the hinterland. All shipping lines will be available at large ICDs and, therefore, offering choices to the customer. The spokes normally located near large consumption or production centres, generating multi-destination single commodity or single customer cargoes. These may be constructed either by end users or container train operators (CTOs). Further, these spokes will be comparatively smaller facilities with low handling costs, and even a significant component of chassis stuffing. The hubs will be large and multi-purpose facilities will develop more as common facilities with assistance from IR and/or respective state governments for acquisition of land. The terminals to be built by CTOs where both containers and rail wagons can be dealt with at mutually agreed terms, for EXIM as well as domestic traffic to be handled.

With more competition coming into the container train business, there appears a greater need for the hub and spoke model in the hinterland. Multiplicity of medium-sized ICDs with competing ownership and management will only make for suboptimal utilisation, and are hence unsustainable.

Let IR optimise utilisation of existing depots and private sidings

With a view to IR winning over general goods in intra-country traffic and transporting it by rail in containers across the country, it may as well build some large terminals as common-user facilities and have a terminal management company operate them.

There is need for many small-sized terminals on the IR network which will tap on-demand train load domestic container business or consolidate traffic even in half train loads. Some of the existing rail freight depots with space available for viable operations can well serve this purpose. IR may let the CTOs use them at reasonable prices so that road traffic is converted to rail. Some of the existing private sidings with spare capacity could also be suitably adapted and optimised for third party container traffic.

While infrastructure projects as a rule remain tardy in terms of implementation, there is need for an integrated and holistic approach for their development. Deficiencies are manifold:

- There is little coordination amongst the various agencies. Additional port capacity for container handling is planned with little consideration to landside logistics. Coordination in infrastructure planning would not only remove bottlenecks, but also avoid overlaps or snags and extra costs. Duly integrated facilities will help reduce transaction costs.
- Tax regimes and recovery procedures as also inter-state transits are cumbersome and time consuming. They need to be simplified and reduced to a one-window/one-time levy across regions so that administrative processes do not hinder free flow of cargo and vehicle movement. The new GST regime will hopefully address these concerns.
- Urban planning does not generally factor in the enormous volumes of goods distribution catering to urban conglomerations. Inadequate road and peripheral infrastructure leads to traffic restrictions and logjams.
- Government agencies do not strongly facilitate proactive and participative dialogue with industry. Blueprints and policy regulations are as a rule a one-sided affair and prone to avoidable trial and error syndrome.
- Information Technology is struggling to find its way into the logistics domain. Acceptance is perhaps not an issue any more, but marriage between IT and domain requirement is nowhere near resolved. Automation in processes is yet only in its infancy. Further progress is also dependent on a certain level of standardisation made more difficult by the high level of fragmentation in the industry. This is a serious drawback and needs to be tackled urgently.
- Spurred by private sector participation, the network of ICDs and CFSs is well poised to expand across the country. Also the share of containerisation is expected to improve in the export/import sector as well as for domestic cargo. With the economy reverting to the high export growth path of around 25 per cent per annum, the EXIM segment is set to be really buoyant. In this steadily increasing multimodal milieu, one critical element requiring sustained attention of industry as well as government is that of human resource development.

For Scanning New Horizons

With a 20-year successful track record in a dynamic multimodal infrastructure development coupled with a compelling institutional mechanism to sustain its growth, CONCOR as the industry leader today is at a crossroads requiring. Good businesses need to constantly reassess their role and re-engineer their plans and strategies. Success sometimes takes its toll. It becomes necessary to do all a business or industry must do to ward off any sloth in the system. Ever new challenges emerging in the global logistics businesses call for timely responses to new aspirations and demands for innovative products and services which must perforce be cost-effective, time-definite, integrated, and global in character. The buoyant industry expects port throughput to reach the 15m TEU mark by 2015–16. This calls for better and more container handling infrastructure at the gateway ports as well as in the hinterlands. Containerised movement of domestic cargo holds a great promise and opportunity.

Tied by an umbilical cord to Indian Railways, CONCOR has inherited infinite strengths and, as it has come of age, it needs to scan new horizons with a new vigour and renewed faith in its creativity, initiative and enterprise; at the same time discarding those activities that may not be of relevance or use in this new phase, and embracing a new paradigm of a new business model. It needs to forge new alliances with global shipping and logistics companies, also to develop multimodal infrastructure and culture in collaboration with many others, for example, in Southeast and South Asia and Africa, sharing experiences, expertise, and business models and expanding investments. Linked by broad gauge (1676 mm) rail network to Dhaka in the east and Lahore in the west, for example, India is well placed to serve as a potential logistics conduit in the South Asian region for Trans Asian Railway system moving containerised cargo through to the Middle East, Central Asia and to Europe.

7.9   Conclusion

Amidst rapid paradigm changes and trade profiles, logistics management has emerged to be an important factor in production and distribution of goods across the globe. As transport platforms are buffeted by increasing competition and rising expectations, technologies evolve, and support systems lend an ever new dimension, emphasis grows for integrated supply chains to deliver advantages in cost as well as speed.

Multimodal transit of containerised cargo is increasingly being developed as a door-to- door facility, for which the dry port plays a pivotal role. Although a somewhat late-starter in developing the multimodal infrastructure, India has taken long strides in its steady and sustainable growth, with a comprehensive framework of systems, procedures and institutions in place. Its rapid growth and commercial success, its multi-dimensional development models may well serve as useful case studies, particularly keen on cost effective and efficient modules with willing involvement of different stakeholders. Development and operation of dry ports in the country has new challenges to face, those of further penetration in the hinterlands along with the need to consolidate and coalesce facilities which have mushroomed in some areas, simultaneously with the need to expeditiously redress some practices which appear to distort the sector and have the potential to debilitate the system.

7.10   References

AT Kearney 2008. *Business and Marketing Plan for Jawaharlal Nehru Port.* [Online] Available at www.atkearney.com.

Container Corporation of India Ltd. *Annual Reports* (various years, including fiscal 2009–10), [Online] Available at http:#www.concorindia.com/index.asp1.

Cushman Wakefield 2008. *Logistics Industry Real Estate's New Powerhouse.* Report. [Online] Available at www.cushwake.com.

Indian Ports Association. *Major Ports of India: A Profile* (various years, including for 2009–2010), 1st floor, South Tower, NBCC Place, Bhishma Pitamah Marg, Lodi Road, New Delhi-110003. [Online] Available at www.ipa.nic.in.

Infrastructure Leasing and Financial Services Ltd (IL&FS) October 2009. Draft MMLP Report, Logistics Park at Rewari on the Dedicated Freight Corridor, IL&FS and Infrastructure Development Corporation Ltd.

Ingram, R. A. H. 1992. *The Development of Inland Clearance Depots, 1952-1992.* International Cargo Handling Coordination Association (ICHCA)-40th Anniversary Review. pp 85-87.

Jawaharlal Nehru Port, [Online] Available at: http://www.jnport.com.

Levinson, M. 2006. *The Box: How the Shipping Container Made the World Smaller and the World Economy Bigger.* Princeton and Oxford: Princeton University Press.

Dobson, H. 1989. Lloyds Maritime Atlas of World Ports and Shipping Places, 16th edition. Alfred Rolington Lloyd's of London Press.

McKinsey and Company 2010. Transforming the Nation's Logistics Industry. [Online] Available at http://www.mckinsey.com/locations/india/mckinseyonindia/pdf/ Logistics_Infrastructure_by2020_fullreport.pdf (August 2012).

Ministry of Commerce and Industry (Department of Commerce), Government of India [Online] Available at: http://commerce.nic.in.

Pricewaterhouse Coopers 2010. *Transportation and Logistics 2030,* Vols. 1-5, [Online] Available at pwc.com.

Rail India Technical and Economic Services Ltd (RITES) 2002. *Study on Preparation of Master Plan for Development* of ICDs/CFSs in India. [Online] Available at www.rites.com.

Rail India Technical and Economic Services Ltd (RITES) 2010. *Total Transport Study.* Report.

United Nations Conference on Trade and Development 2009. *Multimodal Transport Newsletters*, quarterly publications. [Online] Available at www. unctad.org.

United Nations Economic and Social Commission for Asia and the Pacific 2010. Report of the Committee on Transport at its second session. [Online] Available at www.unctad.org.

United Nations Economic and Social Commission for Asia and the Pacific 2010a. *Issues and Challenges in Transport Related to Promoting Regional Connectivity: Transport Policy, Infrastructure, Facilitation and Logistics.* [Online] Available at www.unctad.org.

Chapter 8

Price versus Quality or Quality versus Price at Indian Dry Ports – Cost, Quality and Price – A Visionary vies on Indian Dry Ports

Vaibhav Shah

8.1   Background

The planned capacity expansion at India's major ports and containerised cargo business is all set to respond to the booming demand, and indicates that the biggest beneficiaries of this growth can be the players who provide port-based logistic services, like Container Freight Stations (CFS), Inland Container Depots (ICD) and/Dry Port Operators (DPO). They are an integral part of the logistics chain in relation to the movement of containerised cargo.

Companies like the Container Corporation of India (CONCOR), Gateway Distriparks, Balmer Lawrie, Central Warehousing Corporation, and a few private operators are the dominant players in the field of CFS and ICD business.

Around 30 per cen% of the dry port related traffic is being moved by rail, whereas the other 70 per cen% is being moved by road, mostly to nearby port CFSs and some to dry ports located further inlans.

Under the liberalisation and privatisation strategy the Indian government involves the private sector in infrastructure development, and since 1995 private operators are allowed to developd and operate ICDs (Circular 128/95, Ministry of Commerce). Presently, 240 ICDs/CFSs are operating in India. Around 126 are dry ports, of which almost 30 per cen% are located in the North and Western parts of the country.

Due to fierce competition in the national and international markets Indian producers, merchants or importers are keen to cut downnon the total cost of logistics and to increase the product quality. Transport services providers are focusing on providing integrated, multimodal logistics services for their customers and trying to be part of their production process.

Indian trade has mixed reactions towards quality and price of transport and logistics services. Price appears to be more important than quality for the majority of trade. A perception exists that if a better quality service is available, it should be at a lower price or at least without any additional cost. The main determinant for industry dealing with raw material, bulky shipments, industrial products, intermediate products, agriculture, scrap, raw material etc. is cost, and not the reliability or quality. However, for a furniture trader or importer of high value cargo, price is not more important than the safety and on-time availability of the product. This lead, firstl, to short-cuts or ways which can be seen as unethical practice; second, to low-price demand which will bring down the level of quality, third, to the diversion of traffic from one(DP) to another where capacity resources are under- or over-utilised; an, fourt, and most important, to creating unfair competition.

8.2   Research Objective and Methodology

This chapterexamines the relationship between price and quality in the supply of services at dry ports and its sensitivity to the trade and industries, by investigating the factors or areas that can create a shift in the focus of the dry port industry from a sole cost aspect to a quality aspect to bring about overall benefit to the industry. The identified factors or areas for improvement in the sector should be considered when developing better standards and best practices. This paper aims to show potential and constructive solutions to match the Price-Quality Mix.

An attempt was made to understand and gain insight into each area affecting thePprice-Quality Mix at dry ports by exploring the perspective of the various industry stakeholders based on their focus areas, experiences and expectations.

8.3   Applicable Concept of Price-quality Mix

Customer services are a result of desired activity levels. This implies that each level of services has an associated cost level. In fact, there are many logistics system cost alternatives for each service level, depending on each particular logistics activity mix. Once the service delivery relationship is generally known, it is then possible to match costs with service level (Ballou 1999)

It is a generally observed phenomenon in most economic activities that cost increases at an increasing rate to meet higher customer service levels as they are forced beyond their point of maximum efficiency. On the other side, the service level can also be increased if the cost of each logistics activity mix is audited and controlled. The service level needs to be decided and continuous audits and control are needed to maintain standards. If the service deliveries are below standard, the loss/delay occurred to the logistics activity mix is borne by the customers who remain unsatisfied and gradually shift to alternatives.

While every DPO has its own business model for delivering the service, the underlying activities, quality standards, handling, storing, stuffing and transporting goods are all standardised. However, the price of services is an important factor which decides level of services offered and desired by the Indian customer.

Apart from direct costs, there are other indirect costs which are intangible but have important effect on the services offered by the various activity mixes, which may influence the choice between competing DPs or port CFS. Users of dry ports have suggested improvements in the various tangible services to avoid the intangible cost.

The cycle of tangible and intangible cost is shown below and indicates where in the case of dry ports tangible costs increase for expectations of intangible value.

Even if the direct costs remain constant, the indirect costs are difficult to measure because of their intangible nature and variations in services offered to the different customers.

A connection between the price paid and the services received is based on the quality of service delivery and its indirect cost. For any improvement or additional service an additional price will have to be paid.

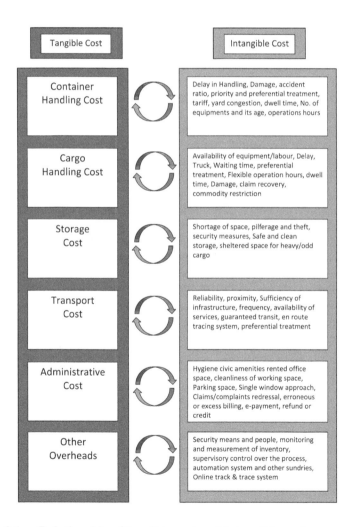

Figure 8.1   Relationship of tangible cost and intangible cost

8.4   Applicability of the Competition Factor

The product of aDdryPport is its "space". Competition builds excess product capacity which is an inherent force to bargain its prices and quality of services. In India, only a few private dry ports were developed before the privatisation of container train operations, but still the boom in the dry port sector is visible. Due to sufficiency of infrastructure and other first mover advantages imperfect competition in India prevails and the dominant market players have a major impact on the price of services. The new entrants thus have a benchmark for their pricing strategy and keep the prices lower in comparison to market trend and existing dominant players. This situation does endanger strategies of investment and cost recovery. A greater parity between competitors would be beneficial to the sector.

In one case the state-owned DP, CONCOR had a competitive advantage over a private DPO for receiving land at a cheaper rate from the state-owned railways proximate to the rail network.

However, it is not the cheaper price of land by its own that gives CONCOR a competitive advantage, but the strategic location of the land, which is not easily available on the market. CONCOR's efforts to develop the land in a way to give boost to the EXIM trade by making the logistics sector more economical. CONCOR is price competitive not because of its cheaper land but its strategic location, services and response to the trade. Acquiring strategic location always gives a competitive advantage of cost and its proximity.

When there is competition, there is comparison. The winning point for private DP operators is dependent on their financing strategy of land in strategic locations, infrastructure and superstructure. Interviews were carried out with local customers of dry ports (i.e. customs agents, shipping agency, exporters, importers, transport service providers), who are the actual users of dry ports, to further investigate the competitiveness and level of services provided by two dry ports having the same captive industries. These ar: CONCOR ICD-Sabarmati, which was established as a fully-fledged facility in the 1990s, an; Thar Dry Port, which was established in 2009, providing basic infrastructure in its first phase. The following section presents the main areas discussed and mentioned by the users of the two dry ports.

8.4.1   Proximity to the Trade and Customer

CONCOR ICD-Sabarmati is located within the city boundaries while the new Thar Dry Port is located approximately 35km from the central part of city. To facilitate and attract traffic, Thar Dry Port arranges vehicles to provide connectivity to/from the city at their own cost in order to overcome the disadvantage in proximity and the additional time and money. They adopted empathy as their quality dimension of service.

8.4.2 TariffsStructure

Thar DP had to adopt the tariff of the incumbent. Shippers offering substantial volume of business are getting few fiscal advantages in terms of lower tariff, discount, credit and convenience to pay.

8.4.3   Train Load Traffic

Since Thar DP only recently started its operation and only small volumes of container traffic are generated, an extreme imbalance between imports and exports traffics exists. Further low volumes do not suffice to justify higher frequencies as long as existing train services are only partly loaded

8.4.4   Congestion

Congestion is the main factor of traffic diversion between the two dry ports. Congestion is a common problem at ICD-Sabarmati, which has been leading to delayed services. Thar DP has the advantage to be able to offer hassle-free flow of their shipments and documents. With The heavy increase in the volume, limitation of space an, being within the city, any other DP could not have managed in such a situation – but CONCOR could do it.

8.4.5   Road Infrastructures

A competitive advantage of Thar DP is its road transport fleet. The operating company owns and runs a number of fleets across the country. Operation costs per tonne/km are low making the company competitive with other existing dry ports in the vicinity i.e. Mundra and Pipavav. Thar DP offers a full range of services including door pick-up/delivery in a single contract. Nevertheless, the interviews with shippers revealed that importers are not happy with the service delivery schedules. Because of late deliveries to importers, shipping line containers remain occupied for longer periods ant the importer has to cover any detention fees, which might lead to a diversion of flows to other dry ports.

8.4.6   Marketing Efforts

Thar DP as the recently established Dry Port is in strong competition with the existing dominant DPO and massive marketing is required to attract cargo volumes and new customers.

8.4.7   Lessons Learned

Even thoug, Thar DP is in its infancy stage, shippers have claimed it to be an equal competitor to existing DPs. The bargaining power of shippers have increased with the additional player in the market and direct comparisons of rates are possible., Since Thar DP is trying to expand its market share to be able to generate economies of scale, the operating company offers significant discounts and credit to the majority of the customers. However, interviewees have pointed out that the DP lacks quality of service, particularly when facing short-term peak demand.

The current competition brings advantage to the users. All trade partners, including exporters and importers, should understand the weakness and threats of market competition to a significantly weaker competitor. A certain level of cooperation would deliver benefits for competition in the long run, but might also bear the risk of too much cooperation. Users have been switching between the ICD-Sabarmati and Thar DP both to seek different advantages such as better attention to their shipments) and to support competition.

8.5   External Threats and Factors Affecting use of Dry Ports

Apart from the above factors, dry ports, in general, are affected by the following external factors

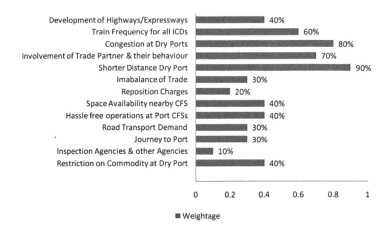

Figure 8.2   External threats for non-usage of dry ports

In spite of the competition from the proliferated port-side CFS operators (JNPT – 30 CFSs, Mundra Port – 14 CFSs, Pipavav Port – 4 CFSs, Kandla Port – 4 CFSs) and numerous road hauliers, which have a 75 % market share of total freight movements at ports, rail transport as a mainstay for DPO can be competitive in terms of costs in comparison to road.

The differences in delivery times and other external factors force shippers to prefer rail instead of rail and dry ports. This not only affects the cost of the shipment but also has repercussions on the environment and road safety. The ability to accurately respond to the dynamics of the global marketplace is no longer months or weeks , hours may decide over the competitiveness of a service and it creates added valueofor the user.

CONCOR stepped up to drive modal shift from road to rail by calculating the opportunity cost of running half rake and empty return journeys at a negligible loss or very less profit margin.

8.6   Priority Areas for Improvement

Price advantage and superior quality, while seemingly having an inverse relation, can actually be complementary. some examples are described in the following section.

8.6.1   *Labour Issues, ReformsAand Relaxation*

Indian dry ports are mostly dependent on significant manual labour. The definition of labour used includes industrial workers, forklift/crane/vehicle operators an, their supervisors, surveyors, and security personnel to work at the dry ports.

Export cargo is delivered in small volumes, which involve various labour gangs to handle them. Manual handling involves a number of risks like cargo damage, damages to packing, pilferage, delay and unethical practices. In India, the level of tolerance fof such incidents is high in comparison to other developed economies.

Dry port operators can play a significant role in this, by reducing the ratio of damage, pilferage and delay. As far as unethical practices are concerned; however, these prevail. This is difficult to avoid as workers are pushing for extra earnings. While giving quality and demand based services it is necessary to not only look at the cost of the services offered but also at the prices paid by customer using these services. The difference in servic-/cos-/price can be used as an indicator of poor measures taken by the operators as well as industry. The government has recently put more efforts into increasing the standard of living of the workers, and it is expected that improvement will be visible in the system.

The labour laws in India are archaic, which, rather than ensuring the welfare of the worker, only leaves the system hostage to adverse action by the statutory agencies on account of so-called violations. There is not sufficient flexibility in the system to allow a margin for minor aberrations in compliance.

It is impossible for firms, contractors or workers to be aware of their rights and obligations when rules and regulations are spread over 45 national and state level acts. The labour acts do in various capacities affect the gamut of logistics activities encompassing material handling, transportation, maintenance and housekeeping, security etc.

In India, the state government under the Minimum Wage Act, 1948 sets the wages at a state level. However, The set minimum wages ar, only a guideline for what a company should pay its worker.yIn addition, the minimum wages are adjusted, revised and increased bi-annually. The present minimum wage structure from 1 April.2009 to 3. September 2009) is presented in the table below.

Table 8.1   Minimum wage scale in India (Indian Rupees, INR)

Wage Structure / Worker Category	Unskilled Worker	Semi-Skilled/ Unskilled Supervisor	Skilled/ Clerical Worker	Highly Skilled worker
Basic wage per day	180	200	220	240
Dearness allowances	44	48	53	57
Total	**224**	**248**	**273**	**297**
PF (12%)	26.88	29.76	32.76	35.64
Admin charges (1.11%)	2.49	2.75	3.03	3.30
ELDI (0.5%)	1.12	1.24	1.37	1.49
ESIC (4.75%)	10.64	11.78	12.97	14.11
Bonus (8.33%)	18.66	20.66	22.74	24.74
Gratuity/terminal benefit (4.81%)	10.77	11.93	13.13	14.29
Leave charge (6.73%)	15.08	16.69	18.37	19.99
Total cost per head as per minimum wages	309.64	342.81	377.37	410.54
Wage per month (26 days' pay) (INR)	**8051**	**8913**	**9812**	**10674**
Wage per month (GBP)	**104.43**	**118.84**	**130.83**	**142.32**

Source: Circular Order No.1/4(3)/2010 LS-II dated 25.03.2010 of CLC(C), New Delh

The difference between the wages of unskilled workers with those of highly-skilled and skilled workers is only £37.89/month and £11.49/month respectively. This affects the motivation level of workers, which in turn impedes performance standards.

A number of points can be addressed to improve the labour situation:

1. The minimum wages could be changed into "living wages" which requires the raising of established government standards, a greater responsibility of the employer towards the workers and effort from the workers themselves by improving the work ethics.
2. Educating workers/employees on rights and responsibilities by the employer to create a change in attitude and behaviour that will justify an improvement of wages.

3. Integrating workers by making them stakeholder in the company's business activities. This can change the attitude towards the company as they feel more integrated.
4. Setting working standards and creating goals to delivering these creates a link between wages, productivity and quality of services.
5. Women are the economic drivers of the country these days. There is no sphere where a woman cannot work; therefore the dry port sector should motivate women to work as contract workers. Contract conditions might even define minimum participation of women in the work force.
6. Choosing of contractors with a well-functioning salary system. For example, workers should be paid through bank accounts, rather than being paid in cash in an envelope every month.
7. Choosing of contractors who have training programmes for workers, not only in the field of work, but also towards the improvement of their standard of living, skill development and their rights.
8. An external audit firm that is expert in human resources may be appointed by the principal employer for audit and control of practices by the contractor, workers and the overall contract system.
9. Beyond these possibilities, the use of cheap labour, most vulnerable to exploitation by contractors, can be reduced by mechanisation, upgrading and automation of dry port operations.

8.6.2   Material Handling

Cargo and packaging damage and loss claims are significant issues faced by shippers. This can be improved if mechanisation is increased as this improves the efficiency of logistics processes and often has positive consequences on costs and quality of services.

The majority of Indian companies have not been progressive in implementing mechanisation and automation. The complementarity of price advantage and superior quality becomes clear under a life-cycle cost analysis. For instance, an expensive piece of machinery or highly mechanised automated maintenance system may initially cost more but in the long run reduces the costs of operations and maintenance. More specifically, the Quantum Series Electronic engines of rubber-tyred gantry (RTG) cranes, while being more expensive, can help to reduce the fuel consumption by up to ten per cent. In parallel, labour requirements can be reduced when using automated systems, not only improving the quality of operations but also yielding a "price advantage" in the long run. Further potentials arise from the use of conveyor belts for stuffing/destuffing containers, electromagnetic devices for handling scrap metal, or scanners for checking consignments when still inside the container. These measures can enable dry port operators to find the least-total-cost alternative while satisfying the requirement of shippers.

8.6.3   *Pilferage*

Occurrences of theft and pilferage of material from the packed consignment at dry ports have been observed. Low levels of control over workers, drivers and crane operators provide opportunities for pilferages, theft and damage of goods and material.

Pilferage is not only happening at the dry ports – the roadside dhabas/eating outlets, parking and rest facilities are among the main spots for pilferage, and theft from parked vehicles and delays in transit cause inconvenience to the free flow of traffic.

Beyond dry ports, CFSs and major public warehousing hubs are also affected by this. This is a grey area where no action has been taken so far and the additional costs of which have so far been tolerated by shippers. Potential solutions might include:

- Strong packaging units that make tearing off difficult.
- Usage of time locks as being used for security of goods in containers.
- Encourage unitisation of loose cargo
- Introduction of standardisation in packaging material and packing standards
- Good surveillance systems, though these come with an additional cost.
- Use of anti-pilferage devices in wagons/trailer which ensure pilferage-free transport of containers are an example of how a simple bar of metal helps keeping commodities free from theft.
- Use of CCTV in warehouses and DP gate complexes woud also help to restrict petty smuggling of commodities.
- Establish controlled rest-rooms, lodging, and eateries along highways.

8.6.4   *Transportation: Accessibility, Speed and Time*

The rail-link or road-link is not only affected by the insufficient supporting infrastructure. Train capacity is defined by Indian railway norms to be between 80 to 90 containers for a full train. Unless and until an operator has a full load at a dry port the train cannot move to the port destination, which significantly affects the service time. This is the opposite in the case of ports where containers are available but due to insufficient capacity supply and turnaround of trains, containers remain at the port for an extended period of time. This also creates congestion at the port. The inefficient turnaround times are caused by inefficient train handling at terminals, due to limitations in engine and crew availability, lack of train stabling lines, rake examination and other administrative delays.

A few suggested ideas, which might help to improve rail efficiency for container freight movements, are presented below.

1. Availability of modern rolling stock and maintenance systems to enable longer run of the stock and hence greater productivity of wagons.

2. Use of longer rakes up to 1500 metres and higher-capacity wagons of 80 to 90mt carrying capacity – as envisaged in the Dedicated Freight Corridor Project.
3. Longer block sections to enable faster transit.
4. Improved last mile rail linkages between ports and DPs to help industry to transport freight more by rail instead of road; e.g. by more loops besides the present CRTs (Container Rail Terminals), DPs or any railside loading facility, and also construction of DPs on existing rail routes for better door-to-ship cargo movement.

Generally, road freight transport is more expensive than rail freight. Various factors contribute to higher transport charges and loss of time in transit such as cost of fuel (up to 50 per cent), high labour charges engaged in un/loading, overloading, road traffic congestion, bad road conditions, toll collection at various points and detention at toll points, city restrictions, etc.

A closer look at the road transport sector reveals the need for strict enforcement of rules to avoid overloading, elimination of unethical practices by trucking companies, replacement of old/outdated with new vehicles, drivers training institute, periodic training on maintenance, road safety, hygiene standards, health hazardous, etc.

The focus on improving state highways and national highways and, to a lesser extent, district and village roads affects the DPs in India, since the majority of DPs are located on the outskirts of cities on rail links that connect to the national or state highway network through district or village roads. The connectivity with national or state highways has traditionally been an issue. ASIDE (Assistance to states for developing export infrastructure and other allied activities) a government scheme has taken a step to assist funding project infrastructure that supports the export promotion. In the programme 25 per cent to a maximum of 50 per cent contribution is generally granted for road development connecting to an economic centre. Beyond building a major impediment is repair and maintenance, especially in monsoon seasons, which affects the quality of DP in terms of its accessibility. It might be considered a contentious action that the government plans, finances and develops national and state highway infrastructure and thus promotes and attracts more road traffic movement, whereas road maintenance is poor for connecting DPs. Road transport being unsafe and non-preferable for multimodal transport worldwide should be demotivated by creating better accessibility to DPs.

At the least, there is a need of an appropriate mechanism whereby the ports with due government support can be encouraged developing end to end connectivity to create a perfect integration of road, rail, gateway ports and other economic centres.

8.6.5 Standardisation

There is also a pressing need for standardisation in dry ports. The key attributes requiring standardisation include:

- Standard design of ICD layout such as location of warehouses, railway lines, stacks etc.
- Standard type of equipments which will help reduce the complexity and cost of maintenance
- Standard systems of the various operational processes and commercial documentation.
- Standard systems of communications
- Standard system of evaluation of processes

Just as the quality of a product can be judged by its conformity to standards and specifications, logistics customer services can be judged by the extent to which service standards are delivered and demands are fulfilled. In short, standardisation aids better comprehension of the processes in the DP, which in return will help to reduce the costs of operation and maintenance while ensuring quality of service.

8.7   Code of Conduct and Regulatory Model

While there is sufficient capacity at the few container port terminals to handle forecasted growth well into the future, concerns exist in relation to the impacts on the longevity of CFSs, dry ports and road traffic to hinterland.

There is an important role for the Government in working with stakeholders to achieve system-wide improvements, such as developing initiatives for improving transport efficiency on the basis of solid evidence with regard to systems and industry practice, a code of conduct, by appointing a governing body for dry ports and the associated transport system.

Apart from a code of conduct, DPOs could seek promotional help from the government which could be setting a dry port policy to establish equal norms for all DPs and minimum standards for quality of services for sustainable growth.

Policy guidelines should be framed, reviewed and updated on bi-annual basis to account for changing markets environments, technology and upgrades around the world.

Numerous ministries, boards, associations govern the various industries, infrastructure development and control trade and services.

To have a broader view and vision an Inter-Ministerial Committee (IMC) for approval of application for setting up of ICD/CFSs exists. Once established, they are governed by the Central Board of Excise and Customs (CBEC) who determines minimum standards for an ICD/CFS (CBEC Circular No. 128/95-cus) from the trade and customs perspectives.

Table 8.2   List of relevant Indian Minstries

- Ministry of Road Transport and Highways
- Ministry of Railways
- Ministry of Commerce
- Indian Port Association
- Director General of Shipping
- Ministry of External Affairs
- Ministry of Law and Justice
- Ministry of Labour and Employment
- Ministry of Shipping
- Ministry of Civil Aviation
- Ministry of Micro, Small and Medium Enterprises
- Ministry of major Individual Commodity
- Export Promotion councils
- Various Federations, etc.

The IMC meets twice a year to dispose of various applications and modifications and new developments. However, the focus should be more on audit and control measures over the practices by existing DPOs, which safeguard the interest of DPOs but damage the services quality and sufficiency of attention to public needs.

While DPs are usually being referred to as being as important as sea ports, they are given different treatment for development and improvements. This author would suggest forming a body or association and a regulatory model that assists, suggests, and guides government in the development of DPs and can represent the sector in the government. The key objective could be to provide a clearer and internationally consistent framework for: improved standards of service, road safety, improved infrastructure, prices, role clarity, standardised codes of practice, minimisation of unfair competitive advantage amongst operators, the promotion of a level playing field for operators, establishing general liability, preventing the breach, better protection for infrastructure and environment, social and welfare aspects, and government support and finance for development and enhancements.

8.7.1   Location Strategy for Dry Ports

Locating ICD facilities throughout the transport network is an important decision that gives form, structure and shape to the entire transport system.

Historically and geographically in India industries are located far from ports i.e. at inland locations. There is no direct relation between the number of dry ports and the number of industrial zones. One dry port can cater to several industrial zones. Many of the early decisions to establish dry ports were based on land availability near the railway network and around city centres and industrial zones.

In rare cases, where industry developed in the absence of a rail network connection, DPs were planned and located as road fed DPs.

Any good location near to industries with good potential, giving advantage over port clearances, can be a strategic location for a developer with the motive of earning profit. However, for overall development in this sector, a strategy or standard should be decided with respect to the location, size, services and maximum usage of existing and future dry ports.

8.8   Chain of Responsibility

There are various legislations under various acts that create an unclear picture of responsibilities. Therefore, a common legislation is to be framed in such a way that it creates a default position of liability for each participant in the transport chain, in order to give a clear understanding of their obligations and required actions.

Codes of conduct and practices are the pillar-stone for standardisation of practices, procedures and documentation, and behaviour of stakeholders and government. A code of conduct can dictate a DP's satisfaction indices and overall satisfaction by the users and operators. Various areas have been identified for improvement, deciding standard practices and bifurcated into three categories:

Table 8.3   Code of conduct for standardisation

Of the Dry Ports	For the Dry Port	By the Government
• Terminal Layout design • Initial Minimum Warehousing and storage • Public amenities • Operations practices • Services Aspects • Safety and Security • People behaviour and Management • Contractual Terms • Pricing • IT Applications • Quality Infrastructure • Maintenance • Grievance redress mechanism	• Location • Documentation • General Safety on Road • Enactment of combined law • Ports operations for dry ports • Rail pricing and traffic management • General liability • Cargo packing compliances • Transport compliances	• Standardise • Frame Policy • Strict Implementation

Apart from the factors discussed in this chapter, financial, technological and knowledge are further factors which impact the comparable standards of dry ports.

8.9   Conclusion

It is evident now that customer service is the net result of the execution of all activities as per target value. Because customer service has a positive effect on turnover, the most appropriate way to approach proper planning is to meet the most economical way possible.

Both quality and prices of service are complex and controversial issues, both technically and conceptually. This is due to provision of infrastructure, economic factors, law and regulations, imperfect competition and the level of state involvement. However, quality minimises cost and wastage. The implementation of quality may not be a burden or an additional cost, but enables a company to get customer's trust and trust means returning use of service. Nevertheless, the cost of improper planning and ineffective implementation can make the service overall more expensive.

The issue of quality and price is leading towards bringing standardisation into the sector. Improper planning, fiscal constraints, status difference, non-standardisation and imbalanced port-dry port integration require government intervention.

This chapter favours government intervention to bring standardisation, for better quality of services and overall development of this sector by not cutting the prices but by cutting the unethical practices among the trade partners in the price war and by re-engineering the conventional competition. Once a certain level of standards is achieved, the way forward is much simpler.

Better facilitation and stringent regulatory environment are the needed most in the current situation to support the present growth. The goal should be to maintain a steady and relatively competitive situation and low pricing with more emphasis on quality and efficiency at Indian dry ports.

8.10   Acknowledgements

I want to thank all people who have been involved in my study. First of all I want to thank TRI/Napier University, who provided opportunity to think and explore about Dry Port Industry in India, and secondly, my organisation Container Corporation of India Ltd (CONCOR) for permitting me to go ahead with this research – especially Mr Amit Kumar Singh, Chief General Manager, North West Region. Many thanks go to all respondents who provided me with feedback, statistics and information.

Finally, I want to give endless thanks to my family and friends, who have supported me on my way to this research.

8.11   References

Amit, N. and Bharat, M. 2005. *The Payment of Wages Act*, 1936, (amended from 09.11.2005). Ahmedabad: SBD Publications, 1–20.

Amit, N. and Bharat, M. 2007. Chapter VI: Working Hours of Adults, in *The Factories Act*, 1948, Second Edition. Ahmedabad: SBD Publications, 36–43.

Amit, N. and Bharat, M. 2007. Chapter IV: Computation and Payment of Wages, Hours or Work and Holidays, in *The Minimum Wage Act*, 1948. Ahmedabad: SBD Publications, 54–67.

Amit, N. 2007. *The Indian Contract Act*, 1872, (Act 9 of 1872), Current Publication, Mumbai (general reference for appointment of contractor).Ballou, R. 1999. *Business Logistics Management*, Fourth Edition. Englewood Cliffs, NJ: Prentice Hall.CONCOR 2010. *The Company, Core Business*. [Online] www.concorindia.com [accessed September 2010].

Department of Commerce, Ministry of Commerce and Industry, Government of India. *Trade Promotion Assistance, Guidelines for Setting-up of ICDs/ CFSs*. Available at http://commerce.nic.in/trade/national_tpa_guidelines.asp [accessed July 2010].Pillai, Smt. Sudha 2008. Initiatives of Ministry of Labour and Employment During Last One Year. Secretary, Ministry of Labour and Employment, Press Information Bureau, Government of India, 1–3.

Public Private Partnership and Competitive Advantage of Indian Dry Ports. Available at www.scribd.com/doc/27184217/Dry-Ports-Public-Private-Partnership [accessed September 2010].

The Contract Labour (Regulation and Abolition) Act, 1970 (Act 37 of 1970), amended by *The Delegated Legislation Provision (amendment) Act*, 2004, Chapter V and VII: Welfare and Health of Contract Labour and Wages. Ahmedabad: SBD Publications, 2005, 31–38.

Visit and telephone Interviews with CHAs, Importers–Exporters, Shipping and Forwarding Agents, CONCOR officers, CFSs at Mundra Port for their perspectives and experience.

Chapter 9

The Construction of Seamless Supply Chain Networks: The Development of Dry Ports in China

Jing Lu and Zheng Chang

9.1 Introduction

Nowadays, ports are striving to perfect their functions as "third-generation ports" and moving ahead to become "fourth-generation ports". Although there is no universally accepted definition for "fourth-generation ports", two trends in the ports sector have been agreed (Xu and Lu 2006). Firstly, competition among ports is transforming into the competition of the supply chains which ports are involved in. At present, the port is a component, rather than an independent point or centre of the transportation chain (Wang 2009). The efficiency of the supply chain depends on the port, and more importantly, all segments linked to the port. Secondly, a port with large and directly connected hinterland has great potential and competitiveness, while on the other hand, a port with a single marketing strategy or which just relies on other countries or regions for feeding will face severe threats.

For Chinese ports, two challenges are faced by port operators: how to find an efficient link to the inland regions to keep the supply chain smooth? How to expand the hinterland and attract more cargo? As for the inland regions, the national strategy to transfer industries to Midwest China brings rather a good opportunity for economic development of the region. The enterprises hope that they can clear customs locally. The inland governments also hope that there will be a platform on which the cargo can transit conveniently, therefore the cities will be closer to international markets and the investment environment will be improved. Driven by the need of coastal ports and inland cities, the dry port concept has emerged as a modern logistics centre located in an inland region with similar functions as those of a coastal port.

According to the report by FDT (2007), a dry port can be described as follows:

> A Dry Port is a port situated in the hinterland servicing an industrial/commercial region connected with one or several ports with rail or road transport and is offering specialised services between the dry port and the overseas destinations. Normally the dry port is container and multimodal oriented and has all logistics services and facilities, which is needed for shipping and forwarding agents in a port.

9.2   Construction Mode of Dry Ports in China

It has only been a few years since the first dry port emerged in China. However, there are now several dry port clusters in China. For example, the northeast dry port cluster which is constructed by Dalian, the mideast cluster which contains 16 provinces and regions led by Tianjin, the southeast cluster built by ports in Jiangsu and Zhejiang province. Generally speaking, three modes of construction are used widely in China.

9.2.1   *Coastal Port as Main Construction Body*

The 16 dry ports built around Tianjin are successful examples of this mode. Actually, the first dry port in China was built by Tianjin in Beijing in 2002. The port of Tianjin occupies only 37 km² of land. However, with the continuous dry port construction, Tianjin can now gain access to 4.5 million km² of hinterland. Cargoes from Beijing, Hebei, Shanxi, Inner Mongolia and Gansu contribute 54.7 per cent of the total commodities value for Tianjin. The "one-stop customs" mode at dry ports reduces costs dramatically, while increasing efficiency for inland shippers. The statistics show that cargoes from mid China save one or two days on transportation and logistics activities. For cargoes from west China, it can reach three or four days. The integrated logistics costs can be reduced by 20 per cent or more (Shi 2009).

9.2.2   *Inland City as Main Construction Body*

Dry port Xi'an represents the second construction mode. In the context of the declining advantage of labour-intensive industries in east China, and the implementation of a strategy of industry transfer towards Mideast China, Shanxi put forward a project called "Xi'an International Port District"; in this way this inland city can get closer to the world and bring itself into full play as the central city of the economic zone of the second Eurasian Continental Bridge. On 1 April 2010, the Bonded Logistics Center was completed, and on 1 June 2010, the container railway central station opened. At present, Xi'an International Port District has signed a letter of intention for dry port construction with Tianjin and is trying to cooperate with Qingdao and Lianyungang.

9.2.3   *Coastal Port and Inland City as Combined Construction Bodies*

In this mode, the coastal port and inland city make a combined effort to build a dry port. Take Dalian as an example. At the second summit of four cities in southeast China in June 2005, Dalian, Changchun, Harbin and Shenyang reached a decision about dry port construction cooperation. Except for Dalian, the other cities are capitals of three northeast provinces, which are also direct hinterlands of Dalian. Northeast China is characterised by abundant resources, historical heavy industries and enormous grain production. But the biggest problem is the underdeveloped foreign trade. The dry port project will certainly push the region into the international markets and boost economic development.

9.3   Effects on Coastal Ports

9.3.1   *A Dry Port is an Important Linkage in the Supply Chain*

The competition among coastal ports can be divided into three levels according to the competitive emphasis. In the first level, ports pursue large scale. In the second level, ports compete for high-quality service, diversified function and low price. Nowadays, the third level of competition, which means competition among the logistics nets and the supply chains ports participate in, is more severe than before. Since the chain depends on every node linked to the ports, the ports should equip themselves with the ability of coordinating and overall planning. At present, ports attach more importance to the construction of seamless international supply chains, rather than blindly focusing on other ports in their range and infrastructure expansion. According to the definition of a dry port, it is just one node in the chain and has close relations with coastal ports; it can collect and distribute cargoes for ports. Thus, the efficiency of a dry port has vital importance for a seamless supply chain (Qin 2010).

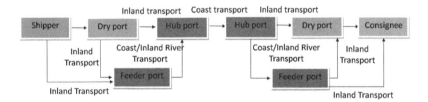

Figure 9.1   Cargo flow in international container transportation

9.3.2   *To Expand Hinterland and Bring Cargoes to the Port*

The ability to attract cargo is significant for the port; more and more ports have given priority to hinterland expansion. This trend is becoming more obvious owing to the financial crisis. Most ports in China have suffered declining cargo and container throughput in 2009. Facing a serious situation, ports are eager to get more cargo supply to maintain the previous high throughput growth. Setting up a dry port means the port can get easier access to the hinterland by offering seaport functions forward to the city. In fact, it is now a rather important and widely-used marketing strategy for seaports to enhance their competitiveness and to overcome this unusual and sudden crisis.

At present, the Pearl River Delta, Yangtze River Delta and Bohai Sea Region have been well developed. Most cities in those areas have a higher level of open economy than other regions. However, the vast inland areas are facing a different situation. Due to their location, resources, and relatively lower political support, provinces and cities have not opened their economies anywhere near the average level. The Chinese government has noticed the potential of inland areas and issued a number of supportive policies for these areas. The strategy aims to boost the development of the central region (which includes the provinces of Shanxi, Henan, Anhui, Hubei, Hunan and Jiangxi) as well as pursuing a develop-the-west strategy (which is designed to stimulate the economy in 12 west provinces and cities such as Gansu, Ningxia, Qinghai, Xinjiang, Sichuan, Chongqing, etc.). With the increasing infrastructure investment; improving investment environment, accelerating processes of industrialisation and urbanisation, the economic growth in inland regions in China will soon surpass eastern areas. It is estimated that the GDP of midland China will increase by ten per cent over the next five years. Transportation demand in these areas will also increase at the same pace. If seaports can grasp the unique opportunity and set up their own logistics platforms in a suitable inland city in the form of a dry port, the hinterland will be effectively expanded, in this way, ports can derive sound development and stand out from the crowd.

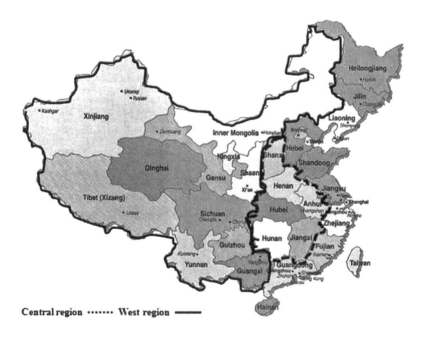

Figure 9.2   Scope of two regional development strategies
Source: Port Dalian (PDA) Co., Ltd

Table 9.1   GDP of central regions in 2009

	Shanxi	Henan	Anhui	Hubei	Hunan	Jiangxi
GDP (billion Yuan)	730	1936.7	1005.3	1283.2	1293.1	758.9
Growth rate (%)	6	10.7	12.9	13.2	13.6	13.1
Ranking	19	5	14	11	10	20

Source: China statistical yearbook 2010

9.3.3   To Release Pressure on the Space and the Collecting and Distributing System

Though Chinese seaports have experienced admirable progress, they are now confronted with the bottleneck of insufficient space. Ports occupy a lot of coastline resource. Therefore, some industries such as tourism might be influenced to some extent. At the same time, with the development of coastal cities and concentration of population, the living community of the city will expand constantly, thus the need for land will increase (Zhong 2010). The consequence of this trend is that there is a great difficulty for ports to get sufficient land for terminals or storage yard expansion. However, a dry port can endow the inland city with the same functions as a seaport and, in this way some logistics activities, such as storage, devanning and consolidation can be carried out in inland areas. Consequently, the pressure of space needed during port expansion can be eased enormously.

On the other hand, a dry port can alleviate the pressure on the collecting/ distributing system tension of the port. Before the "dry port" concept appeared, the transportation between inland areas and ports was poorly organised. Most vehicles had low loading rates which caused wastages and a lack of organisation in cargo collecting and distributing. In addition, most cargoes were carried by road. Traffic congestion, insufficient carrying capacity and environmental pollution are severe problems that need to be addressed urgently. A dry port can change the transportation pattern dramatically. Cargoes can be collected in a dry port, hence goods in large quantities can use railway as the mode of transport. In this way, the transportation cost is lower while service quality is higher; the ability to collect and distribute cargoes smoothly is improved. Figure 9.3 shows conventional transport patterns and the change triggered by dry ports.

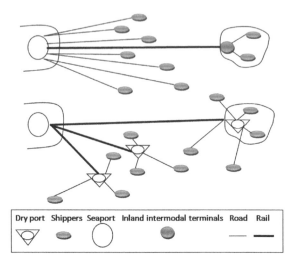

Figure 9.3   Comparison of conventional transport and dry port concept
Source: Modified from Roso 2009, p. 303

9.4   Driving Effect on Opening Economic Development

9.4.1   Dry Ports Contribute to Coordinated and Balanced Development in China

The Chinese economy is characterised by regional disequilibrium and imbalance. Southeast areas along the coast have been economically developed, while inland regions in the northwest have a lagging behind economically (Lu 2010). Recognising the great potential of resources of the region, China has implemented several strategies to stimulate economic growth. These policies will provide unprecedented opportunities for inland regions. Transportation needs keep increasing in these years. As the origins of goods moving from coastal to inland regions, it is critical for the inland cities to have more functions in order to provide shippers with convenient transportation (Zhang 2009). A dry port is a platform to fill the regional resource gap and can shorten the geographic distance to markets. In consequence, the inland region is able to open up to the outside world, and the economic imbalance in China can be eliminated.

9.4.2 To Improve the Investment Climate and Activate the Local Economy

An inland dry port with customs and similar-to-port functions, as well as an international logistics channel, is definitely the driving force for opening of a city's economy. It is also an important platform to get access to international markets and improve the investment climate. Therefore, a dry port can assist the city to attract more foreign investment to inland regions and take an active part in undertaking industry transfer in east China. In the aspect of market motivation, cost motivation and resource motivation, boosting foreign investment will increase import and export trade in the region. In conclusion, a functional dry port is the trigger for improving the environment improvement and the vitality of the economy.

9.4.3　To Reduce Transaction Costs and Increase Customs Efficiency

The customs function of a dry port can improve the efficiency of inspections, thus dramatically reducing transaction costs. In a dry port, inspection, quarantine and other supporting customs services are centralised and electronic declaration is widely used, so the procedures are quite simple. At the same time, the information is shared between customs of coastal ports and dry ports. After a one-time clearance and inspection, the import or export cargoes will be supervised by customs during transportation; this will simplify the inspection procedure and speed clearance procedures. Take dry port Lanzhou as an example, it is estimated that companies can settle their exchange accounts seven days in advance and get tax drawback 30 days earlier at least. At the same time, transportation costs will be decreased by 50 per cent, however as the possible damage caused by double inspections and loadings can be avoided (Zhu 2009).

9.5　Empirical Analysis of Dalian

9.5.1　Fierce Competition Faced by Dalian

According to the Liaoning Ports Planning, Dalian is the hub port in the cluster. However, Dalian is facing enormous pressure from three nearby ports, Yingkou, Jinzhou and Dandong. The throughput growth of these three ports has exceeded Dalian. In recent years, Dalian has lost the competitiveness and market share on some traditional cargoes such as grain, steel, ore and timber. At present, Dalian focuses on international containers, bulk cargo carried by rail and local cargo.

Although Dalian's throughput has been in first place for many years, the up and coming Yingkou has put great pressure on the port and challenged its relative monopoly status. This trend can be seen in the changing market share in cargo throughput. Figure 5.1 depicts the decline in Dalian's share in the Liaoning port cluster since the beginning of the twentieth century. In 2001, nearly 70 per cent of cargoes transported through Liaoning ports were handled by Dalian, while in last year, the share dropped to 49.1 per cent. In contrast, the market share of Yingkou has increased rapidly from 15.3 per cent to 31.7 per cent. The throughput of Yingkou in 2001 was merely 22.68mt, which was one seventh of Dalian. In 2007, for the first time Yingkou succeeded in being one of the eleven Chinese ports with cargo throughput over 100mt. Yingkou's throughput has grown at an average annual rate of 29 per cent, which is higher than the national average.

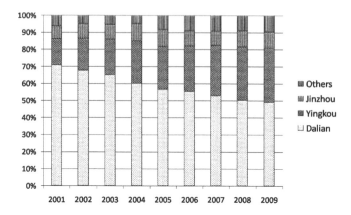

Figure 9.4   Change in throughput in the Liaoning port cluster
Source: Liaoning statistical yearbook 2001–2010

As for container throughput, the change in market share is similar to cargo throughput. Dalian's container throughput share has been falling ceaselessly. In 2009, only 56.3 per cent of containers were handled by Dalian, compared with 81.7 per cent in 2001. Instead, Yingkou has taken more container throughput. Its market share has doubled in the past ten years, increasing from 14.2 per cent to 31.2 per cent. It is also worth noting that Jinzhou performed very well. In 2001, only three per cent of all containers were transported through Jinzhou, but in 2009, the number jumped to nine per cent.

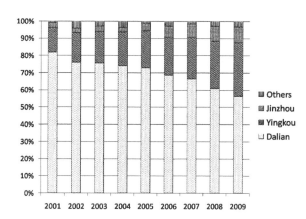

Figure 9.5   Change in container throughput shares in the Liaoning port cluster
Source: Liaoning statistical yearbook 2001–2010

As a matter of fact, compared with other ports, Dalian has a clear advantage in terms of infrastructure, service, collection and distribution network, as well as the economic scale, financial services, new technology and human resources of the city. This embarrassing situation of a declining market share was caused by a disadvantageous location and being a long distance away from its hinterland. Yingkou is 190km nearer to the northeast hinterland by road than Dalian. Under this circumstance, Dalian put forward the construction of a dry port, with the purpose of perfecting integrated services for inland clients and gaining more cargo resources from its hinterland. Dalian has made great efforts to improve the rail–sea intermodal transportation system and to enhance the connection with railway lines through dry ports. Dalian is attempted to maintain its competitiveness in the inland areas in the means of effective, reasonable investment and cooperation.

9.5.2   Dry Ports Layout

Dalian and Liaoning international transportation Co. Ltd built Shenyang goods yard in July 2003. From then on, Dalian has been increasing its investment in inland node construction. Today, Dalian has shaped a dry port layout which covers northeast inland areas. The distribution can be summarised as "three logistics centres and three inland node zones".

Central stations: Dalian, Shenyang and Harbin   On 20 September 2006, Dalian and China Railway Container Transportation Co. Ltd established the "Dalian Railway United International Container Transportation Co. Ltd". According to the cooperation agreement, three railway container central stations would be built in Dalian, Shenyang and Harbin. With the stations' completed, the capacity and service of Dalian's rail–sea intermodal transportation system of Dalian will be greatly improved. This would guarantee the influence and brand effect of Dalian in the hinterland.

Dry ports layout: Three zones and several nodes   On the basis of the logistics network planning in northeast China, Dalian divides its dry ports network into three zones, which take Shenyang, Changchun and Harbin as the respective centres. Specifically, the three zones include ten cities: Shenyang, Changchun, Jilin, Harbin, Qiqihar, Daqing, Suifen and Manzhouli will contribute as dry ports, while Tongliao, Yanji, Muling and Jiansanjiang are planned as container yards.

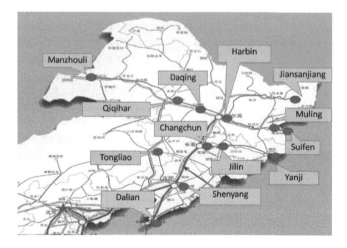

Figure 9.6   Dry ports layout
Source: collected by the author

a. *Dry port Shenyang* Dry port Shenyang takes the responsibility of extending the port function of Dalian to the Shenyang economic circle, and includes the six relatively well-developed middle cities of Liaoning province. It has multiple functions such as inland customs, international multi-modal container transportation, freight forwarders supplying, etc.; it is also entailed with integrated logistics services on the base of tradition terminal yard functions. The freight volume of Shenyang via rail–sea intermodal transportation has been the highest of all for the past seven years.

b. *Dry port Jilin and Changchun* The first time that railway freight yards were modified according to the demand of customs services in Jilin and Changchun. These two dry ports in the Jilin province have given important guidance for container dry port construction in other inland cities in China. In order to attract grain from the two cities, Dalian independently invented the technique of transporting grain via container. Using advanced technology, easily-accessible dry ports and convenient container trains, Dalian has formed a modern grain transportation system. In June 2010, container throughput of Changchun dry port reached to 33,742 TEU, double the amount at the same time of 2009. Jilin's throughput came to 15,970 TEU, an increase of 35 per cent over the previous year.

c. *Dry port Daqing* The dry port in Daqing is anticipated to come into operation at the end of 2010. It is adjacent to large-scale petroleum enterprises such as Petro China and Sinopec Corp, which can bring nearly 1mt of container cargoes. Its locational advantage enables the dry port to save transportation cost effectively and gain access to nearby cities such as Qiqihar more easily.

d. *Muling logistics terminal yard* Muling is located in the golden triangle of northeast Asia and also on the critical gallery for exports to Russia. The city has an intensive transport network which can extend in all directions. The project started in October 2009 and will be finished in November 2010. With dry port Muling, a speedy, effective and convenient transportation channel connecting Dalian with northeast inland regions will be created.

9.5.3   Container Train Operations

The successful operation of container trains in northeast China began in 1997, when Dalian and Harbin were connected, marking the beginning of rail–sea intermodal transportation services between the port of Dalian and inland regions. Since then, Dalian has been devoting itself to constructing an inland collecting and distributing system. At present, Dalian is able to offer transportation agent services in the whole northeast region. There are six block container trains which travel to Shenyang, Harbin, Changchun, Yanji, Manzhouli and Tongliao, with 50 trains going there and back each week.

In 2007, Dalian opened the first international route which starts from Dalian, by way of Manzhouli, Zabaikalye, Chita, and ends in Moscow. Taking advantage of the international train route, Dalian successfully attracted more cargoes from Japan, South Korea and south China; the brand effect of rail– sea intermodal transportation service having greatly expanded. Today, Dalian endeavours to make itself the new bridgehead of the Eurasian Continental Bridge.

Figure 9.7   Container trains station layout
Source Port Dalian (PDA) Co., Ltd

The advanced container train network makes the layout of dry ports and inland nodes more reasonable, and has improved Dalian's the collecting and distributing system. From 2000 to 2009, Dalian maintained in first place among all coastal ports as shown in table 10.2 in terms of container throughput completed via rail–sea intermodal transportation.

Table 9.2   Dalian's container throughput via rail–sea intermodal transportation

	2000	2001	2002	2003	2004	2005	2006	2007	2008	2009
Total container throughput (1000 TEU)	1,011	1,217	1,352	1,670	2,202	2,350	3,212	3,810	4,500	4,550
Growth rate (%)	37	20	11	24	32	21	21	19	18	1.1
Containers transported by rail–sea	44	68	82	197	182	136	159	180	235	253
Growth rate (%)	-	54.5	20.1	141	−8.5	−26	17.1	14	31	8.0

Source: China statistical yearbook 2001–2010

9.5.4   Operational Processes of a Dry Port

The operational processes of inland intermodal transport are as follows:

a. A shipper sends the cargoes to a logistics central station by road.
b. Cargoes are handled in the station (e.g. devanning, storage, maintenance, customs declaration, customs clearance and inspection) and then loaded on a train.
c. The train is marshalled and transports the cargoes transported to the yard of the container terminal.
d. Cargoes are carried to the wharf apron via transportation facilities.
e. Cargoes are loaded on a ship and transported to the destination.

Viewing the cargo flow between a dry port and Dalian, it can clearly be seen that a dry port greatly simplifies customs procedures. All the inland import and export trade can be realised in a dry port. A dry port has the same functions as a seaport, except for the loading and unloading operation; it can offer the specialised services of a seaport such as chartering, booking, signing the bill of lading, as well as customs functions such as declaring and inspection. Inland regions are able to enjoy a one-stop service which means "one declaration, one inspection and one release" by sharing the same customs platform as Dalian. In consequence, the logistics cost is lower than before.

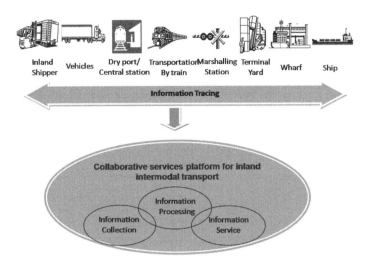

Inland Shipper	Vehicles	Dry port/ Central station	Transportation By train	Marshalling Station	Terminal Yard	Wharf	Ship

Figure 9.8   Brief flow chart of inland intermodal transportation

9.5.5   Challenges and Difficulties

Although the dry port construction strategy has shown its positive and effective impacts for the port of Dalian, there are still some practical problems during the operation.

a. Inland government should play far more important role   Though three dry ports were built by Dalian and local governments working together, the governments merely played a coordinating role in the process. For example, the authorities of Changchun and Jilin just led the cooperation for Dalian and the railway administration and offered a platform for communication. As for the infrastructure, most was invested by Dalian. On the one hand, this will cause severe capital problems to the port. What is worse is that the cooperation between the port and the government is rather loose and not thoroughly guaranteed. Tonghua, another city of Jilin province, is now the dry port of Dandong. There is a chance that Tonghua will take cargo away from the two dry ports, in the end, and from Dalian. Therefore, Dalian should try to encourage the local government to take more responsibilities, perhaps in terms of investing in facilities or building a logistics park together, or to persuade the government to institute policies to encourage cargoes to go through local dry ports. For the port, this will release capital pressure and enable them to keep the cargoes; while for the government, this will make the dry port layout more reasonable, make better use of resources and promote the regional economy more effectively.

b. Cooperation among the involved parties should be strengthened   Dry port construction is system engineering (Yang 2006); its efficiency depends on the cooperation of several actors such as governments, customs, port authorities, shipping companies and shippers. However, some local governments and authorities may take an uncooperative attitude just out of subjective and short-term interest. For example, sometimes the customs authorities are reluctant to take the possible risk caused by the "one-stop" customs mode of dry port. This will absolutely impede dry port development.

c. Huge investment and low direct payback give port operator great capital pressure   A dry port is a capital-intensive infrastructure. It may take millions of dollars to set up the logistics platform in inland cities (Zou 2009). Until now, the Dalian port authority has invested 6.5 million dollars in the logistics centre in Muling and 10 million in Dalian central station. It is estimated that 9.5 million dollars will be invested in Harbin central station. In 2009, the total rent for the three dry ports in Changchun, Jilin and Shenyang accounted for 5.8 million dollars and is predicted to rise to 5.9 million dollars in 2010. Huge initial investment and rent have exerted tremendous financial pressure on the Dalian port Container Company. Therefore, the government should take this into consideration when issuing relating supportive policies to attract social capital in dry port construction.

9.6   Conclusion

The main objective of this chapter is to summarise port development in China. The "dry port" concept has just appeared in China over the last couple of years. However, the economic strategies which focus on the vast inland with tremendous resources and great potential have triggered enthusiasm for dry port construction. Port authorities have noticed that traditional port competition has evolved into competition among the supply chains the ports are involved in. As another link in the chain, a dry port is of great importance for the construction of seamless supply chains. For seaports, a dry port can bring a port's functions forward into an inland city and guarantee the supply of cargo. Meanwhile, it can ease a port's demand for land for expansion by moving logistics activities to the city. While for inland regions, the diverse functions a dry port possesses will attract more investment and promote local economy. These will coordinate and balance Chinese economic development.

A case study analysis about dry ports set up by the port of Dalian concludes that there are quite a few challenges faced by Dalian. The analysis provides some valuable operational implications for port authorities with the planning of dry port construction planning. Port operators should have closer cooperation with local governments and try their best to strengthen the relationship among other actors such as shipping companies, customs, and railway operators. The formation of seamless supply chains should be based on the joint efforts for every linkage and segment along the chain.

9.7   References

FDT 2007. *Feasibility Study on the Network Operation of Hinterland Hubs (Dry Port Concept) to Improve and Modernize Ports' Connection to the Hinterland and to Improve Networking.* Integrating Logistics Center Networks in the Baltic Sea Region Project.

Gong, P. 2010. The Significance of Construction of Xi'an Dry Port. *Value Engineering*, 11, 235–236.

Lu, L. 2010. Promote Inland Economic Development in the Aspect of Supply Chain. *Port Economy*, 3, 32–34.

Qin, Y. 2010. The Realization of Seamless Trade by means of Dry Ports Construction. *China Business and Trade*, 08, 230–231.

Roso, V. 2009. Emergence and Significance of Dry Ports: The Case of the Port of Göteborg. *World Review of Intermodal Transportation Research*, 2(4), 296–310.

Shi, L. 2009. Analysis on the Strategies of Tianjin's Dry Port Construction. *Logistics Sci-Tech*, 10, 19–21.

Tan, K. 2009. Construct Dry Ports Cluster and Raise Core Competitiveness of Guzhou. *Chinese Water Transportation*, 7, 20–21.

Wang, G. 2009. Research on Inland Dry Ports Construction and Development Mode. *Port Economy*, 3, 27–30.

Xu, W. and Lu, M. 2006. Dry Port in the Role of Port Development. *Water Transport Management*, 9(28), 8–9.

Yang, R. 2006. Analyze on Inland Dry Port Construction in China. *Port Economy*, 5, 53.

Zhang, L. 2009. *Research on Interactive Development between International Shipping Center and Inland Port.* Dalian: Dalian Maritime University.

Zhong, F. 2010. Analysis of Dry Ports Construction and Development in China. *The Business Circulate*, 18, 23–24.

Zhu, T. 2009. How to Promote Open Economy Development of Interior Area by Dry Port: A Case of Lanzhou Dry Port Project. *China Business and Market*, 4, 62–65.

Zou, Y. 2009. Discussion on Dry Ports Development in China. *Chinese Water Transportation*, 12, 18–19.

PART IV
The Americas

Chapter 10

Observations on the Potential for Dry Port Terminal Developments in the United States

Bruce Lambert, Chad Miller, Libby Ogard and Ben Ritchey

10.1   Introduction

Over the past few years, several books on various aspects of globalisation have been published. *The World is Flat: A Brief History of the Twenty-first Century*, discussed the movement towards globalisation, one of which is the efficient transportation networks that link markets together (Friedman 2005). In Rivoli's *The Travels of a T-Shirt in the Global Economy: An Economist Examines Markets, Power, and Politics of World Trade*, the global economy not only influences production and consumption regions, but by extension, the global political balance (Rivoli 2005). Other books have elaborated on this theme, but the implications are clear: the world is "flat" and interconnected, as increased telecommunications and reliable transportation networks have, and will, transform access to markets, information, and economic growth.

There have been many groups, especially in the US, promoting the expansion of US exports. President Obama signed the National Export Initiative (NEI), which focuses on the service needs of promoting small- and medium-size companies to engage in international trade (oftentimes exports are promoted through lending programs, assistance programs, and trade missions). However, there are other components that are supported by the transportation activities, which include access to equipment, intermodal facilities, and other critical support elements that promote international trade activity. While such discussions suggest international trade is not important to the US, it is quite the contrary. International trade has exploded in the US, with total trade volumes (in value basis) growing quite strongly over the past 20 years. This is especially true for the growth of containerised shipments, which despite the recession of the past few years, saw dramatic increases continue from the very modest levels of the 1970s and 1980s (Figure 10.1). Despite the current decline, most forecasts assume that containerised cargo has recovered from the recent recession and that the growth in containerised trade will average 4–5 per cent over the next twenty years.

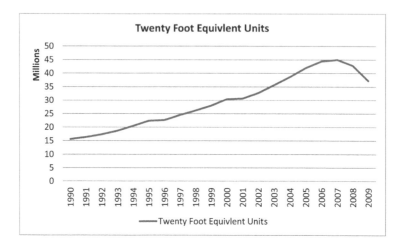

**Figure 10.1   Total US containerised trade activity in Twenty-foot Equivalent
Units (TEUs)**

Source: American Association of Port Authorities

The growth of containerisation has led to different markets for US business,
and realignment of traditional trade lanes. Ports are now an interchangeable link
in the system, not a separate component of a transportation activity, as companies
plan and execute global supply chains. Information about the products must be
viable throughout the process. Intermodal transportation connects trucks, trains,
ships and warehouses via intermediaries to move goods between manufactures
and consumers. These complicated relationships tend to optimize each partner in
the supply chain, which generally requires freight density and balance in order to
be profitable.

Given the recent focus on further supporting US exports, the role of linking
inland hinterlands to ports becomes even more important to support economic
activity. In this paper, the role of dry ports as facilities that link containerisable
activities with either import or export shipments is presented. (It should be noted
that often dry ports are called inland ports in the US, which means they can
provide services similar to a coastal port – customs, consolidation, etc., without
geographically being within a coastal port area. Inland ports also generally include
warehousing and other transportation services.) The author also assumes that a dry
port will have the following characteristics: it may be a multijurisdictional corridor
(crossing over national, state, or regional political boundaries) but the actual project
is normally tied to a specific location; it is multimodal (while there may be one
mode that captures the largest market share, other cargo options exist as well); it
may be part of a larger corridor or trade lane; and it serves containerisable cargoes.
To support existing inland hinterland connections, there are many things that
have to be considered, from both an operational side and a planning perspective.

Oftentimes, there is joint marketing along a dry port corridor, as well as other related promotional activities. In Table 10.1, a list of characteristics of various dry ports indicates that the term itself can vary from the large gateway terminals in the Chicago area to small rail ramps scattered across the US.

Table 10.1   Elements that may be involved in a dry port operation

Characteristic	Agent or Service Required
Ownership	Private Railroad, Private Terminal, Public Development Agency Port Authority
Modal Services to Gateway Port	Railroad, trucking
Ancillary Investment in Dry Ports Area	Warehousing, Distribution facilities, Transloading
Cargo Services in Dry port Area	Customs, Cargo Consolidation, Inspection (Not always available)
Construction and Maintenance in Dry port Area	Public (Most have various degree of public sector grants, road construction, utilities, etc.) Private
Marketing and Promotion of Corridor	Public Authorities, Private Companies, Regional Chambers of Commerce
Service Patterns along Dry port Corridor	Daily, Weekly or other standard operating schedules
Cargo Destination	Mix international/domestic cargo, international only
Cargo Type	Container cargoes

Source: Authors

In sum, the term dry port can easily be expanded to cover any terminal tied to a specific port region. This paper simply defines a dry port into a basic intermodal freight facility that has a dedicated terminus at a coastal port, with rail services being the primary mode linking the dry port to the coastal port. Despite the range in scale and operations, in all cases, dry ports seek to attract and service regional shippers, but the use of the dry port term is often used without any clear distinction regarding any minimum configuration or operation structure. While the success of any dry port (or intermodal freight facility) is dependent upon train density and traffic volume, there exist a host of factors discussed below that are shaping the future of dry ports (and freight intermodal terminals) in the United States.

10.2   The Inland Hinterland and Port Connection

Years ago, the lack of adequate landside transportation (tunnels, bridges, roadways) to inland regions limited a shipper's choice in determining which ports to use. Even today, the region adjacent to a given port is considered "local or captive," as the cost of using another port and shipping the product may be prohibitive.

For example, a shipper based in Washington State would not ship their containers through a California port only to reposition these containers back to Seattle if a service to Seattle was available at a competitive cost in time and money. (In some ways, the outsourcing of transportation has resulted to shippers not necessarily being aware of the shipment's actual routing.) In contrast, discretionary cargo can be handled at multiple ports. If the cargo can be handled efficiently at more than one port, the shipper and/or carrier can choose which port best suits their unique needs, as is the case with shipments destined to Chicago from Asia. In the US, shippers can route the cargo through the US West Coast or East Coast, depending upon service and costs requirements. The balance of local versus discretionary cargo has shifted over the past twenty years. Since the 1980s, with the adoption of intermodalism, the deregulation of the US transportation system, and innovations in logistics management have opened up new markets for various ports and inland markets although operational challenges still exist, such as chassis pooling.

This section will discuss three things: how ports connect to inland hinterlands, the role of intermodal rail freight in the United States, and the current Port and Intermodal Traffic Patterns.

10.2.1   How do Ports Connect to Inland Markets?

A port's competitive hinterland may be defined by the markets that the port serves (ports depend upon serving a consumption area, a production area, or a transfer function). Before containerisation, ports had natural inland hinterlands that generated and received cargo. Today, intermodalism, combined with global supply chains, allows shippers more options in selecting production–distribution activities, including how these supply chains utilise gateway ports. Shippers primarily route cargo with the goal of maximising the returns on their transportation dollar (this applies equally to intermodal or general cargo operations). The chief concerns in selecting a particular route involve price and service, but other factors are equally important. Some of these factors are: what types of vessels call at a particular port; what additional services are available in a port area; what rail or inland connections are available and their relative capacity, and how the port fits into existing distribution patterns. However, the flexibility shippers now enjoy means to some extent ports have lost some of their monopoly power. Today, ports are part of a larger system, and the lengthening of supply chains (due to globalisation, access to markets, financing, etc.) implies that the relative importance of each node becomes diminished by the growth of alternative routings between markets exist when shippers are balancing production and consumption flows.

The focus on developing reliable networks becomes more important in allowing ports to develop larger market areas, and attracting additional business to their facilities. Making ports more attractive by providing additional market access is even more critical as larger containerised vessels are entering service. The arrival of larger container ships has led to many changes at container ports, which can result in altering day-to-day operations. The physical plant at a given terminal may be unable to accommodate the new vessels, resulting in new configurations of berths, turning basins, and other structures. Similarly, the inbound gate areas that initially was designed for processing one set of container traffic, must now handle higher volumes and the increased surge associated with larger containerised vessels. Corresponding related increases spread throughout the entire supporting infrastructure necessary to operate port facilities. Shipping channels, marshalling areas, highways, and rail lines are often not readily capable of handling the higher demands placed on them by the increased throughput generated by the newest and evolving generations of container ships (Abt and Lambert 2006).

   This leads to important considerations about access to land to develop terminal capacity. Dry ports, with an emphasis on collecting services outside of the traditional port boundary, are seen as a way to develop markets by operating facilities outside of the traditional port area and providing scheduled services to and from the port. As such, the ability of a corridor to develop land outside of the port area may provide benefits to both parties, especially as a port's effective terminal capacity becomes geographically spread out to other facilities, thus possibly reducing localised truck moves in a port area while allowing for additional market penetration. This is not necessarily a new concept. Port and inland hinterland connections often developed in the past for dedicated bulk or break-bulk services (such as grain, coal or fruits). It is the evolution of interrelated/interoperable dry ports providing ongoing, scheduled common carriage services that provides unique opportunities for many different agents to benefit from improved services and costs. This is different from the traditional port centric development where the port, with its nearby supporting services and infrastructure, served as a self-contained enterprise zone, where cargo was transloaded, shipped, processed, etc., continued to depend on others working around the hinterland and network deficiencies at a port or terminal area.

10.2.2   *Rail Intermodalism is a Growing Part of the US Transportation System*

The US transportation system is a complex interrelationship of ports, highways, railroads, waterways, and pipelines. According to the Organisation for Economic Co-operation and Development (OECD), the United States is the largest market for transportation services, largely due to its diverse geography, range, and the breadth of cargo (Organisation for Economic Co-operation and Development 2009). For the US, domestic cargo remains the largest single transportation market, as imports and exports each accommodate a significant value of tonnage. While the US has traditionally been engaged in international trade, the advent of containerisation with changes in domestic transportation operations resulted in tremendous changes in the North American network flows.

   Since the 1980s, railroad reforms (including the Staggers Act) made it possible for railroads to engage in expanded services to international shippers and other logistics firms. The result is that rail intermodal traffic, consisting of both domestic and international containerised freight (Container-On-Flat Car (COFC) or Trailer-On-Flat Car (TOFC)), is becoming one of the leading commodity groups carried on the US rail network. The majority of the containers move through the US in the dedicated double stock COFC network, which allows for a doubling rail capacity when compared to TOFC operation. The increasing share of intermodal rail movements has been supported by trucking companies and other integrated carriers as railroads have begun offering more consistent and reliable services, although there are challenges to balancing equipment and reducing costs to remain competitive against trucking in many markets.

During this same period, firms benefitted from increased telecommunications capabilities, customs modernization, and more reliable transportation networks that squeezed inefficiencies out of the logistics system (i.e., lowered costs) while simultaneously expanding the use of transportation through improved supply chain visibility. While intermodalism did not develop because of investment by shippers, these groups have profited from the additional service and options intermodalism provides them. The costs associated with developing transportation activities are partially absorbed by the transportation industry, as evidenced by the historically poor rates of returns on transportation assets, overcapacity issues, etc. Thus, shippers want consistent speed and reliability but expect these services at competitive prices. Over time, the cost of logistics as a share of Gross Domestic Product has fallen to roughly 17 per cent in the early 1980s to roughly 8 per cent of the US economy in 2009 as the result of these efficiency gains (Council of Supply Chain Management Professionals 2010). This has occurred at the same time the growth in total US spending on transportation for all modes exceeded US$ 1.1 trillion.

**Figure 10.2   Picture of double stock Container-On-Flat Car (COFC) in the
                 Chicago area**

Source: Bruce Lambert, Private Collection

Figure 10.3   Picture of single Trailer-On-Flat Car (TOFC) configuration in the Chicago area

Source: Bruce Lambert, Private Collection

10.2.3   Current Port and Intermodal Traffic Patterns

The US has a tremendous port system, as highlighted by the recent "America's Container Ports: Linking Markets at Home and Abroad" (US Department of Transportation, Research and Innovative Technology Administration, Bureau of Transportation Statistics 2011). The largest container ports are Los Angeles and Long Beach, located in Southern California, which serve as the intermodal gateway from the Pacific to the eastern United States. Of the top ten ports, most both serve large local or regional markets and provide regional transportation gateway services.

As shown in Figure 10.4, there are coastal differences regarding the number of ports, volumes and traffic patterns. The Pacific Coast is more reliant upon handling intermodal shipments arriving in the US (with the notable exception of Oakland, which serves more as an outbound load centre for vessels returning to the Far East). The Port of Houston and most Gulf Coast ports have larger export shipments of manufactured and consumer products. Along the East Coast, the overall container trade is more balanced to all markets (not just Asian cargoes), as imports arrive in the major markets of New York, but Norfolk, Savannah, and Charleston have heavily invested in attracting inbound Asian traffic to their regions. The irony is that there are relatively sparse container markets along the West Coast, which supported market consolidation and cargo densities, while the more dense port system along the eastern US has led to fierce competitive positions regarding traffic growth. This also means that the eastern US ports are more reliant upon

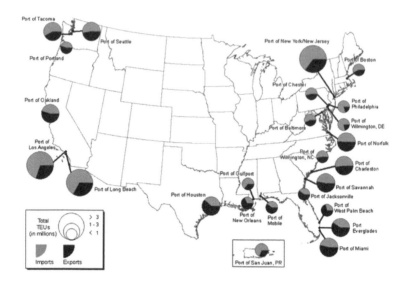

Figure 10.4   Top 25 container ports in the United States, 2009
Source: US Department of Transportation, Research and Innovative Technology
Administration, Bureau of Transportation Statistics, 2011

trucking as a share of total movement to inland markets, especially as their captive
markets are fairly close to the coast when considering Asian cargoes.

When one considers the flow of intermodal traffic, the linkages to ports
becomes more apparent. The largest intermodal flows originate arrive from Asia
on the West Coast and are destined to Chicago before heading to other markets
in the eastern US The largest of these flows involves shipments between Los
Angeles/Long Beach and Chicago, but other major flows include traffic from
Oakland and the Seattle/Tacoma area. There are plans to increase other intermodal
corridors, such as expanding intermodal services from the Port of New York/New
Jersey, and the recently completed Heartland Corridor (Norfolk, Columbus, and
Chicago). When considering rail intermodal terminals, it is easy to see how the US
rail intermodal market for international cargoes flows along two main East–West
patterns. The first pattern, largely based in the West Coast involves flows into the
Chicago area where intermodal cargo is interlined or transferred to other railroads.
The second, of which the primary flow passes through Southern California,
involves shipments to and from Texas and other rail interchange terminals along
the Mississippi River. The question regarding the development of these flows is
that most of the US economy is based in the denser eastern US, while the western
US networks are sparser but with large urban centres. As such, the question of
developing or supporting dry ports must recognise the respective densities moving
between various ports and potential inland markets in the eastern US, which will
largely remain the central driver to the US economy.

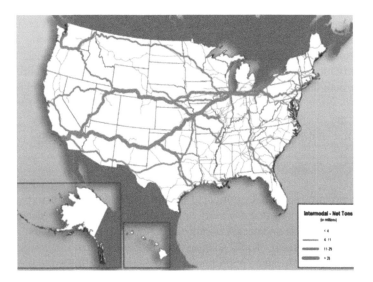

Figure 10.5   Tonnage on trailer-on-flat car and container-on-flat car intermodal moves, 2006

Source: US Department of Transportation, Federal Railroad Administration, November 2010, from Federal Highway Administration, Freight Facts and Figures 2010

While most economists agree trade can promote economic growth and development, for the US, the opportunity to capture these benefits may be dampened by the inability of the port industry to handle future volumes. A 2003 report issued by the US Chamber of Commerce (Trade and Transportation: A Study of North American Port and Intermodal Systems) suggested that, by the year 2020, the volume of international trade will have nearly doubled from the base year of 1998. The report determined that most of the major container ports in the US would face critical issues of capacity, with limited ability to respond or handle increased traffic. While all gateways that handle international trade are seeing increased congestion, more ports are facing difficulties in obtaining land for new facilities, developing on-dock or near dock rail intermodal facilities, and congestion on the highways that service a port area. These hard infrastructure constraints, and their resulting drag on terminal productivity, may ripple throughout the supply chain, potentially dampening economic growth or allowing other ports or services to capture market share.

10.3   Why Develop Dry Ports?

The use of dry ports allows ports to access additional markets in a more timely or reliable manner if full trainload densities can attract reliable services. While generating additional traffic for the port, the linkage itself can develop additional traffic volumes and better service, perhaps leading to lower per unit costs as well as increased use of the system for other cargoes that may be moving along the same corridor. For example, the large movement of intermodal containers from Los Angeles to Chicago shifted in most of the cargo in that trade lane onto railroads from trucks for both international and domestic cargoes. There are three main reasons to develop dry ports: to develop port traffic; to provide new opportunities for regional shippers; and to alleviate intercity highway congestion along a corridor (although dry ports may not actually alleviate local congestion around the facility).

10.3.1   Port Traffic Expansion

While dry ports are considered critical to expand a port's economic activity, it is also important to the markets that dry ports can service but this is largely determined in the US by the location and access to railroad networks. A recent article highlighted that intermodal access is clearly shaping the competitive position among various ports in the US, but the same opportunities can be equally afforded to the shippers at the originating or terminating inland markets (McSwain 2011).

For most port authorities, the pressure to grow traffic volumes that support local economic development remains the primary purpose in their public charter. However, the changing world of vessel operating strategies means that fewer ports may be in a position to evolve into megaports, further intensifying the demands for ports to provide competitive inland services to attract vessel traffic to their facility. The potential alignment of shipping services in the aftermath of the Panama Canal expansion has clearly demonstrated that ports must not only be able to handle the anticipated draft of the larger vessels, but also provide the necessary terminal space and inland connectivity that results from larger surges of containers during a port call.

10.3.2   New Cargoes to a Hinterland Cargo

The Network Appalachian Study identified the inability to access international markets as a critical deterrent to supporting economic growth in their thirteen state regions (Appalachian Regional Commission 2009). The study identified that access to global markets, largely through inaccessible or costly transportation networks, limited the ability of the region's exporters to engage successfully in international markets. The study recommended continuing to build the Appalachian Development Highway System, but also creating a series of intermodal "corridors of commerce," that can attract and build upon regional traffic flows. The third strategy recommendation was to develop a series of new inland ports (dry ports) based on successful regional inland ports. There are two main ways that these flows may change hinterland cargo volumes: changing gateway flows and business park development.

After the lockout of the West Coast dockworkers in 2003, shippers began diversifying their traffic through various gateways to both increase system redundancy and improve access to various regional markets. This has resulted in some market share shifts for Asian containerised trade from the West Coast to the eastern US There are many who consider the expansion of the Panama Canal a potential game changer regarding shifting additional traffic from the west coast to the eastern US through the new deployment of larger, more efficient ships calling at south-eastern regional ports (roughly 70 per cent of the cargo that transits the Panama Canal has an origin or destination in an eastern US port.) The Panama Canal may result in a shift of Asian imports to markets readily accessible by truck from Eastern US ports, but also reaching to other inland markets. There exists some discussion concerning the ability of eastern ports to provide intermodal services to Chicago, Kansas City, and other major markets in the Midwest regarding Asian imports transiting through the Panama Canal. (Most people see the battle for intermodalism within the US to focus on the inbound cargoes from Asia, a discussion that often includes comments on inventory carrying costs, bunker fuel, and vessel schedules, which partially skews the broader discussion on rail intermodalism access for export shipments and container availability.)

Business parks and business clusters are two of the focus areas that economic developers consider with inland terminal development. In various site selection surveys, transportation access is rated as a very high concern for business, and a KPMG study stated that 90 per cent of all the surveyed businesses indicated that transportation directly influences their business activities (KPMG International 2009). Transportation needs are critical for any site under consideration for foreign direct investment as these firms are more likely to engage in both import and export shipments than a comparable domestic firm. Encouraging local agencies to work with carriers and transportation departments early in a project would ensure that new sites may achieve their full potential regarding traffic volumes and operations. This may move logistics to the top of the economic development list for the region, but improved coordination between economic development or State commerce agencies and transportation providers could avoid straining transportation budgets with last-second requests for infrastructure projects.

For example, the south-eastern US has been successful in attracting new automotive production (Lambert and Miller, forthcoming). The focus for economic development has been centred on business clusters around a major facility, as evidenced by the location of the larger suppliers. Furthermore, for the BMW plant in Figure 10.5, their supply chain depends upon two ports that handle engines and other parts for their assembly line. Normally, when considering foreign direct investment, economic developers often ignore the larger supply chain considerations when trying to secure a major project. As such, neighbouring areas may benefit from additional businesses locating nearby to support the plant but also may suffer the drawback of additional traffic on their infrastructure, resulting in additional congestion.

Oftentimes, business parks may be linked to a specific foreign trade zone. Foreign Trade Zones (FTZs) represent secured areas under US Customs authorisation and supervision that allow firms to delay or manage customs duties. Operating in secured locations, the cargo has physically entered the US, but remains outside of the US for customs purposes. In addition to supporting commercial activities, they do provide public benefits, such as encouraging US exports, attracting offshore investments, and providing employment and economic development opportunities. (Wilbur Smith Associates 2005). As FTZs must be within or adjacent to an approved Customs and Border Protection (CBP) gateway (airport, border crossing or port), the potential exists to support dry port/business centre development programs through FTZ's can authorise subzones, which are normally private businesses engaged in export related activities.

10.3.3   Alleviate Highway Congestion

There is a final push for developing dry ports in the United States to alleviate intercity highway congestion, both regionally and nationally. Ports generate large volumes of drayage movements, as containers are moved to various places around the port for transloading, inspection, delivery, etc. The role of localised port traffic has resulted in several port cities trying to restrict or impose various operational strategies to reduce congestion and emissions.

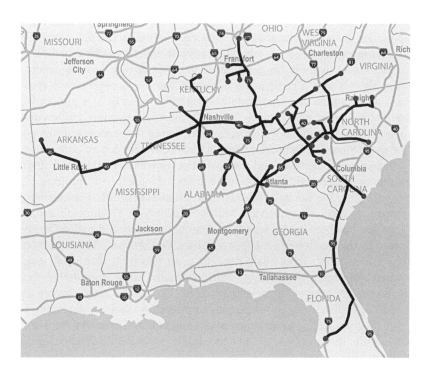

Figure 10.6   The estimated interstate flows from its major suppliers to BMW's greenville plant

Source: Lambert, B., and Miller, C. (forthcoming)

The Federal Highway Administration's (FHWA) Freight Analysis Framework attempted to estimate the national system performance. The darkest lines in Figure 10.7 show high volume trucking corridors, which are largely uncongested, in 2007. By examining traffic patterns based on reported volumes and system characteristics, FHWA estimates congestion based on traffic levels reaching 75 per cent of anticipated design levels. At this point, traffic speeds begin to decline, resulting in system delays. At levels of 90 per cent of anticipated design criteria, traffic levels see strong declines in average speeds. Already, all classes of urban areas are experiencing traffic congestion, a fact that has grown over the past twenty years, according to the 2010 Annual Urban Mobility Report (Texas Transportation Institute 2010).

According to FHWA, in 2007, congestion was a significant factor in major metropolitan areas, with a few local corridors handling large volumes of traffic (Figure 10.7). By 2040, the predicted congestion not only spreads to more of the urban areas, but many of the major interstate corridors will experience reoccurring traffic delays (Figure 10.8). While the majority of the freight shipments on the US system are domestic, the resulting flows will also influence international traffic moving through the same networks.

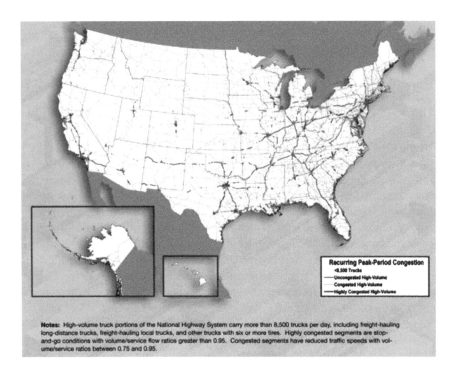

Figure 10.7   Estimated daily highway congestion, 2007

Source: US Department of Transportation, Federal Highway Administration,2010

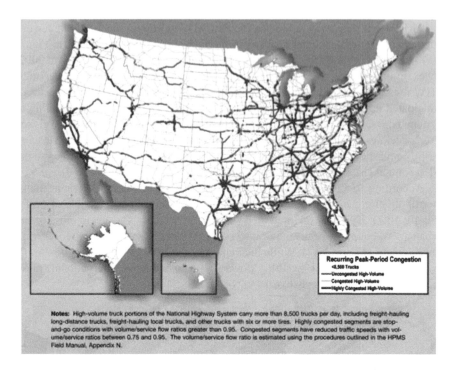

Figure 10.8   Estimated daily congestion volume in the United States in 2040
Source: US Department of Transportation, Federal Highway Administration 2010

Concerns over congestion remain a challenge, as lost productivity results in many costs to the system. There are differences concerning the costs of delays upon truck movements, ranging from delay costs of US$26.70 (US Department of Transportation, Federal Highway Administration 2009) to total costs for a trucking company of US$83.68 (American Transportation Research Institute 2008). In the future, given the relatively slow construction of new capacity on the US Interstate system, it is expected that more segments will see daily peak period congestion. In response, carriers and shippers have considered various options, such as locating distribution facilities in largely suburban or rural areas that have lower land and operating costs, as well as limited congestion, and moving more deliveries to off-peak hours and other modes. As such, the large dense movements are reloaded onto smaller lots to better schedule deliveries around congested networks. Clearly, moving international cargoes to and from regional distribution facilities would provide some incentives for firms to locate around business parks, as is the current case in the Columbus, Ohio, area and the Rickenbacker facility (Battelle Technology Partnership Practice 2007).

Such a strategy possesses merit, as increasing numbers of firms are mentioning that transportation access is becoming a growing concern for site development.

Clearly, the development of logistics parks, FTZs and other economic clusters are becoming more important in developing regional businesses, and shippers are increasing wanting to have modal options when locating in a business park setting. These parks require reliable services, often with rail access, to attract firms to locate there. Furthermore, by consolidating international and domestic cargoes on other modal networks, the service may generate sufficient cargoes to be attractive to various service providers and perhaps produce traffic flow changes, such as backhauls to develop additional cargoes.

10.4   Institutional Frameworks for Operating Dry Ports

Given the importance of developing dry ports, how does one go about creating such relationships? There are various agents involved in the development of a dry port. These may include:

- *Port authorities and public entities.* These are traditional port operators but may also include other public entities, such as regional economic development groups or authorities. Normally, these are created by a state or local government body with specific jurisdictional and operating guidelines. A recent study by PIANC Working Group 31 discussed the governance of inland maritime ports, but many of the same guidelines apply to coastal ports as well (International Navigation Congress 2010).
- *Local Economic Chambers of Commerce/Economic Development Group.* The economic development role that a Chamber plays and the mix of business members in a Chamber makes for a logical place to organise logistics (freight) players for an inland port (dry port). Also, the mix of players reflects the variety of businesses involved in logistics ranging from the normal players (shippers, carriers, third party logistics providers) to indirect players (development, government, legal, finance, academic, software/technology firms).
- *Transportation Companies.* Drayage operations have come under increased scrutiny in various ports, concerning localised traffic, emissions, and the condition of the trucks themselves. In the US, railroads are the major drivers in intermodal terminal development and access. While some dry ports are served by multiple railroads, most are captive to a single railroad company because of rail competitive issues.
- *Shippers.* Shippers are becoming increasingly aware of the importance of freight productivity, especially as it relates to knowing with certainty when the product will arrive and being able to track the container's location. However, shippers do not necessarily want to hear problems related to service, so their involvement tends to be limited in most cases, unless they are a captive shipper to a certain trade lane. Many shippers have had bad experiences with railroads in the past twenty years when railroads were reducing costs and

services to match their revenues. However in the mid 2000's, the railroads experienced a turn in business interests focusing on intermodal terminals to survive in a more competitive world. Some shippers have long memories and war stories with significant financial consequences, which make "rethinking railroads" sometime difficult internally while others are seeing rail strategies as critical to their long supply chains.

- *External Pressures.* Transportation decisions are often overlaid with a host of other decision makers, involved parties, etc., that can severely shape or restrict various transportation projects. These can include the ability to access roads for a dry port (such as financing rail or highway interchanges) as well as land use decisions that can potentially change a port's ability to generate productivity gains through building or expanding intermodal facilities. In many cases, the local community determines the path of economic growth based on other concerns related to local transportation or land use patterns not necessarily in accordance with the economic benefits of increased traffic through a freight terminal.

10.4.1   Market Information

For any relationship to succeed there must be economic returns or benefits for all parties, as well as a flow of information, collaboration and other details. This section will discuss some elements that should be considered when developing a dry port, such as understanding market information, terminal configurations, financing considerations, and managing expectations.

Oftentimes, economic development entities focus on developing their region without understanding the role of transportation (largely rail freight in the US) or its related linkages to markets, etc. In most situations, there generally are not single entities responsible for freight movements at ports and borders. At what level should the private sector discussion occur: at the port, the railroad, the shipper, or the carrier? What is the public sector role: Is it defined by the port, customs authorities, local departments of transportation, and other federal or state agencies? In some areas, each of these groups needs to have some level of information on activity at the port or border crossing, but at different times and scales. For example, the private sector is examining events within the context of a few days – such as securing the necessary documents, or the trucks, to move the cargo to or from a border crossing. The public sector is either responding to goods already in transition or very long range planning activities. The specific information needed for these different levels are not the same, although common features do exist. It should also be noted that generally data is not collected for planning purposes but rather to assess rates and fees against the cargo by customs or to track cargo by the private sector. This may result in the data not necessarily being available or consistent for all users.

There are limited datasets for national level planning of dry ports within the public domain in the US, such as the Freight Analysis Framework or the Railway bill sample. The lack of analytical information often requires local developers to fund their own analysis. In addition, market studies are often complicated by the lack of operational data, such as rates, equipment availability, and other costs that could determine a project's success.

Oftentimes, this analysis is complicated by the inability of data sharing with various public or private agencies regarding specific local data (either for competitive reasons or simply because it does not exist). This question of balancing planning roles has been well documented in various *National Cooperative Freight Research Program* (*NCFRP*) studies, including Report 1: Public and Private Sector Interdependence in Freight Transportation Markets and Report 2: Institutional Arrangements for Freight Transportation Systems (Transportation Research Board, National Cooperative Freight Research Program 2009). However, even if perfect market information is available, the ability to forecast future traffic can be complicated. Not only should the forecast assess the various market factors at play, but also potential changes to the system, such as business patterns and service requirements.

In sum, market research questions may include: What is the competition doing? Will the railroad participate? Is there enough demand to support a balance train operation that will generate a train a day to a single destination? Will the train network support a new train per day? What equipment is needed to carry the cargo? What are the shippers and carriers? Finally, what are the prevalent transportation costs, rates, and service needs for these particular shipments to be competitive? The ability to identify what specific industries can or will use a dry port, in both directions to balance equipment flows, lie at the heart of any market study, to identify not only volumes, but also shipper needs.

10.4.2   Terminal Configurations

The following section demonstrates the nature of intermodal terminals in the US (Lambert and Ogard, forthcoming).

Gateway Terminals are normally located at the end of a rail carrier's network and represent the large terminals that receive multiple trains daily. These facilities can also be along various interchange locations (e.g. Chicago and Memphis) where the trains interchange cargo between various railroads. These facilities often handle high volumes of containers (both international and domestic) with a focus on quickly moving the cargo through the facility (either by truck or on another railroad) to prevent terminal congestion. Containers leaving these terminals are typically moving to the next train, not to customers. As such, their cargo volumes are determined by large cargo flows (such as at a port or major hub area) which generate high traffic counts.

Intermediate Terminals serve as points along a railroad network. These terminals exist to secure additional cargoes for the mainline scheduled intermodal trains, which can provide competitive services against trucking or other railroads. The majority of the cargo serviced by these terminals will be domestic, but some international containers will be moved through these terminals. The success of these terminals depends largely upon the ability to secure traffic from local markets.

A paper ramp represents a service operated by a railroad that does not necessarily have the cargo volumes to support dedicated services or where equipment imbalances exist. These terminals are normally served by trucks to other regional or satellite terminals that can consolidate cargo volumes for movement along the railroad while providing equipment (chassis and containers) for local companies to load intermodal cargoes. To the customer the railroad service pick up time and cut off for outbound shipments "on paper" looks just like a rail served operation.

Given the high costs of operating railroads in the United States, the ability to develop ongoing traffic densities is critical for successful dry port operations to exist. (For rail to be competitive, largely against trucking, a railroad needs to have either length of haul or density.) In some cases, a service may start with a paper ramp or through an existing intermediate ramp, and growth is incremental over existing services where appropriate. Furthermore, the blending of both domestic and international cargoes moving along the same corridors and destinations can also help develop densities. Finally, if cargo volumes grow, other cargo traffic may be attracted to the area around the terminal, resulting in additional traffic for the railroad.

10.4.3 Financing Considerations

Normally, any terminal property remains expensive to develop, but port facilities are especially expensive, oftentimes because they are located in urban areas that may have higher land use values. The US Public Port Development Expenditure Report estimates that US public ports spent US\$ 1.1 billion on facilities, infrastructure, dredging and security in 2006. Of this, roughly one third was spent on container facilities, while spending on infrastructure (roads, utilities, etc.) amounted to only seven per cent (US Department of Transportation, Maritime Administration 2009). Despite the high cost of infrastructure, ports still spent over US\$ 18 million on rail improvements on or near port facilities in addition to on-dock rail or other dedicated costs for specific terminals in the US in 2006.

However, for most inland terminals, the terminal itself lies outside of the traditional port boundary, and port authorities are largely unable to bond or generate funds for these projects. (Most ports are also very heavily indebted because of the high costs of developing maritime facilities.) For example, Long Beach and Los Angeles could participate in developing the Alameda Corridor but are unable to invest significantly in Alameda Corridor East, (expanding additional rail capacity east of the port area). Some State port authorities have the ability to bond and develop related port economic development zones throughout the state, much as Virginia did to develop Front Royal or the way West Virginia will fund the construction of the Prichard Intermodal terminal.

For the US, several intermodal terminals are funded by state or municipal authorities not related to the coastal port. The Port of Memphis has been working on a rail connection with the Canadian National Railroad to link Memphis to

Prince Rupert. The Huntsville (Alabama) International Airport has developed a rail intermodal facility to serve local shippers. The Rickenbacker Global Logistics Park is managed by the Columbus Regional Airport Authority.

However, the majority of the dry port terminals will largely be built and/or operated by private railroad operators. As such, these large terminals are often a component of a railroads operational network and have been criticised as not being responsive to local market needs in the development of new intermodal terminals. However, any railroad project must be coordinated with other rail activity on the same network, which can limit equipment availability and/or the train's ability to deliver expected intermodal service.

While the project may be operated by a private railroad, the financing of dry ports has largely been the result of public private investments. The railroads industry was successful in receiving TIGERI and TIGERII Grant awards, but sometimes this grants called for the state to build specific terminals for railroad access in a corridor. Oftentimes, a public agency (state, city or regional port/development authority) is expected to put up a sizable portion of the initial development costs. The mixing public and private funding is not without challenges, as some projects have been plagued by concerns that the railroad was too heavy handed or a lack of transparency regarding the use of public funding to develop these projects.

10.4.4   Managing Expectations

Most people simply believe if you build something, someone will use it, and that trade will flow easily through the facility. This is somewhat misleading. We can point to many facilities throughout the world that are under-utilised for a number of reasons. There are problems with planning for infrastructure; generally, it is either a "lumpy" large-scale investment or incremental improvements to an existing structure or operational improvements. This approach of "either/or" investment underlines the fact that infrastructure improvements require the spending of funds – funds normally secured by either the public sector planning process or by private sector investment. (In some cases, funding is hard to secure from the public sector, so the private sector has become viewed as a way to fund infrastructure through public–private partnerships.) The need exists to service growing traffic, but physical constraints may limit the ability to expand existing terminals quickly to handle unexpected cargo growth. In other cases, the potential challenge of locally active participation by other groups concerning current and future use of specific gateways may change a location's competitive position. While physical or process upgrades can improve the flow of cargo through a facility, in some cases local groups have sought to reduce dry port projects because of concerns over traffic congestion, noise, and air emissions. The "rise" of active local opposition to large projects remains a potential issue, as "Not in My Back Yard" challenges has resulted in many large freight projects either scaling down anticipated construction, or even preventing new projects from going forward.

10.5   Future of Dry Ports in the US

Traditionally, transportation projects were considered to represent a public good, where public sector money was spent on constructing and maintaining the infrastructure necessary to support the movement of people and cargoes. Part of the challenge in developing dry ports results from the very haphazard manner in which systems planning is done by various sectors. Normally, there is a "champion" who is strongly advances the program. This champion can be driven by political solutions (public) and not necessarily market based goals (private sector), which may complicate the actual execution of developing a successful project. Furthermore, the location of the champion, either from an inland market or from a port will influence their ability to build coalitions (Wilmsmeier, Monios, Lambert 2010.). Within the US, a variety of different groups has responsibility for financing, planning, designing, constructing, and operating the nation's infrastructure. Over time, public sector practitioners (and institutions) set up guidelines and recommendations based upon identifying their specific needs (and/ or data availability). Oftentimes, these design or operational needs were very detailed in some areas related to a research project, while other elements remain partially considered or not even estimated at all. (Private sector activities use other measures to estimate their return on investment without consideration of the public estimates concerning benefit cost ratios, such as return on investment.)

The future of transportation infrastructure improvement in the United States will depend upon the emerging debate over transportation planning/policy/ operations regarding not only the financing for both new and existing infrastructure, but done in a manner that is both environmentally sustainable, safe, and fiscally responsible. At the heart of this deliberation will be a focus on the determination of the new role for both the public and private sectors, which will largely consist of understanding freight needs by the industry, system capital and operating costs, and aligning regional transportation networks to support these movements. Within the US, the public investment is limited to "hard infrastructure", such as roadways, locks and dams, and navigation channels. The private sector is responsible for all equipment, crewing and operational costs, and for the railroads, this includes trackage, terminals and signalling.

Clearly, there are benefits to investing in railroad corridors to alleviate intercity highway congestion, as well as reducing emissions, etc., that should make railroads a more attractive component of national freight movements (Commission for Environmental Cooperation 2011). Given the need to improve rail systems to handle increasing traffic (both domestic and international), the railroad industry has been very aggressive in reinvesting in rail capacity but also tunnels, grade crossings, and other operational improvements (technology, signally, etc.). While there have been several projects that involve financial partnership with federal funds (such as from the TIGER Discretionary Grants Program) the extent of those funds represent a small portion of the current investment portfolio and current budget discussions suggest this trend will not change in the future.

One caveat exists: as dry ports become expected drivers for regional economic development, the potential for overbuilding dry port terminals (mostly from political pressures) remains a potential threat to their individual success. Furthermore, the potential mismatch of trying to develop facilities through political will without a clear railroad partner and effective market based decisions can (and has) resulted in many "white elephants".

10.6   Conclusion

The development of dry ports promises many benefits, such as connecting to global markets, economic development, and opportunities for regional/local businesses. However, what are the benefits of discussing dry ports? The Latin American Trade and Transportation Study recommended that several factors should be addressed to improve the mobility of international cargoes (Wilbur Smith Associates 2001). These recommendations, highlighted in Table 10.2, suggest that a single solution will not address critical freight infrastructure and operational needs. Regional strategies, of which the connection of ports to hinterlands, may require considering dry ports as a specific component of a broader network.

Table 10.2   Recommended freight solutions to address the movement of international cargoes on regional, multijurisdictional networks

Utilise Existing Infrastructure	Attention to Connectors
Add Physical Infrastructure	Encourage Technology
Increase Operating Throughput	Integrate Information
Adopt Corridor Approach for Investing	ITS Applications
Develop Agile Freight Operations	Increase Public Awareness
Improve Clearance at Gateways	Further Institutional Relationships
Improve Freight Profile	Foster Partnerships, Private/Public

Source: Authors

The listing above may be distilled into four topics: Institutions, Information, Inter-operability and Infrastructure. Institutions provide the framework for developing, operating, and maintaining the infrastructure systems. These represent the local and federal authorities who develop, use, or operate the port systems, railroads, and highway networks. Information about the system is critical, as people need to understand its use, its potential, and their relationship to other players (decision makers) regarding both operational matters and information on the economic value the system provides. Interoperability refers to the integration of various transportation elements, such as services, schedules and equipment availability – normally the actual elements that move the cargo. Finally, infrastructure includes not only physical inventories of structures, and terminal configurations, but also construction of new facilities or the maintenance of existing facilities. Among the four, most work tends to focus on the "infrastructure" question. How do we build the system? What investment is required, and how will new technologies improve the system and its use?

There appears to be a roadmap for developing dry ports. If an inland market (or port) is lacking a partner, the hinterland and network deficiencies limit it from actively supporting traffic densities. If the players can secure or guarantee traffic within the port/dry port corridor, the focus shifts to the development or linkage to a terminal/ intermodal transfer service. If the relationship is profitable or provides beneficial activities, all expect additional movement around the facility or along the corridor. However, these relationships can be fickle, especially if not all agents understand the commitment necessary to support current traffic or attract new business.

Dry ports represent a potential focus for future freight mobility, but it will require a longer term approach to build, operate, and maintain these facilities, as improving transportation means so much more than it did fifty, twenty, or even ten years ago, incorporating concerns over flexibility, improving operations, and positioning for handling uncertain traffic forecasts. Today, dry ports must balance the needs of finding or expanding available space along a rail network, securing cargoes to maintain market based services, and securing public funds as necessary to develop the initial infrastructure. Clearly, dry ports, with a focus on supporting trade activities, can provide one area of enhancing regional and national movements, as well as economic development, but many challenges exist regarding the development of a national network of dry ports within the US.

10.6   References

American Trucking Research Institute 2008. *An Analysis of the Operational Costs of Trucking*.

Abt, K., and Lambert, B. 2006. *The Development of Larger Container Vessels at Port Facilities of the Eastern US Container Ports: Changing Port Operations and Infrastructure Investment*. Lisbon, Portugal: PIANC Congress 2006.

American Transportation Research Institute. 2008. *An Analysis of the Operational Costs of Trucking*. Arlington, Va. American Association of Port Authorities 2010. *Port Industry Statistics* [online]. Available at: http://www.aapa-ports.org/ [accessed 9 June 2010].

Appalachian Regional Commission 2009. *Network Appalachia: Access to Global Opportunity*. Washington, DC.

Battelle Technology Partnership Practice 2007. *Central Ohio's Logistics Roadmap for Complete Columbus*.

Commission for Environmental Cooperation 2011. *Destination Sustainability: Reducing Greenhouse Gas Emissions from Freight Transportation in North America*. Montreal (Quebec), Canada.

Council of Supply Chain Management Professionals 2010. 21st Annual *State of Logistics Report*. Wahsington, D.C.

Friedman, T.L. 2005. *The World is Flat: A Brief History of the Twenty-first Century*. New York: Farrar, Straus and Giroux.

International Navigation Congress 2010. *Governance Organisation and Management of River Ports*. InCom report 110, 2010. 10-14 May 2010, Liverpool, UK

KPMG International 2009. *Bridging the Global Infrastructure Gap: Views from the Executive Suite*.

Lambert, B., and Miller, C. (forthcoming). *The Southern Auto Industry: Driving Forward*.

Lambert, B., and Ogard, L. (forthcoming). *Considerations for Rural Economic Logistics Parks*.

McSwain, C. 2011. *The Efficiency Quotient: How Ships, Trains and Trucks Converge Says a Lot*. Site Selection.

Organisation for Economic Co-operation and Development 2009. *Trends in the Transport Sector 1970–2007*. OECD Publishing

Rivoli, P. 2005. *The Travels of a T-Shirt in the Global Economy: An Economist Examines the Markets, Power, and Politics of World Trade*. Hoboken, NJ: Wiley.

Texas Transportation Institute 2010. *Annual Urban Mobility Report*.

Transportation Research Board, National Cooperative Freight Research Program 2009. *Report 2: Institutional Arrangements for Freight Transportation Systems*.

US Department of Transportation, Federal Highway Administration 2008. *Freight Story 2008*. Washington, DC.

US Department of Transportation, Federal Highway Administration 2009. *Freight Facts and Figures 2009*. Washington, DC.

US Department of Transportation, Federal Highway Administration 2010. *Freight Facts and Figures 2010*. Washington, DC.

US Department of Transportation, Maritime Administration 2009. *US Public Port Development Expenditure Report (FYs 2006 and 2007–2011)*. Washington, DC.

US Department of Transportation, Research and Innovative Technology Administration, Bureau of Transportation Statistics 2011. *America's Container Ports: Linking Markets at Home and Abroad*. Washington, DC.

Wilbur Smith Associates 2001. *Latin American Trade and Transportation Studies*. Wilbur Smith Associates (http://www.ittsresearch.org/adobe/LATTS1-final-report.pdf)

Wilbur Smith Associates 2005. *Latin American Trade and Transportation Studies (LATTS II) Foreign Trade Zone Briefing Paper*. Wilbur Smith Associates

Wilmsmeier, G.; Monios, J.; Lambert, B. 2010. *Observations On The Regulation Of 'Dry Ports' By National Governments*, Paper Presented at the International Association of Maritime Economists Annual Meeting, Portugal, Lisbon. 7–9 July 2010.

Chapter 11
Intermodal Freight Corridor Development in The United States

Jason Monios[1] and Bruce Lambert

11.1   Introduction

While some degree of multimodalism (the use of multiple modes) has existed throughout history, intermodalism in the modern era is a relatively new phenomenon. During the mid-1980s, carriers operating in the transpacific trade were suffering from excessive tonnage and lower rates. To increase its cargo volumes, American President Lines (APL) formed the first transcontinental double-stack train services, recognising that an intermodal routing provided a ten-day service advantage over an all-water service through the Panama Canal to New York. While the transit time was important, APL also offered more services to the shipper as the customer could receive a single through bill of lading while knowing that APL had committed service schedules to deliver the cargo.

The growth of discretionary cargoes allowed APL and other shipping lines to expand their capacity in the transpacific. By using larger, faster ships, a carrier could offer a fixed, weekly sailing schedule, while the additional capacity reduced per unit costs. With the double-stack train, these new services were competitive because they increased the amount of revenue that each unit train could generate, provided a shipper with a single through bill of lading and lowered the net cost of inland transportation.

Traditionally, intermodalism in North America referred to discretionary cargoes destined for areas east of the Rockies, but that arrived along the West Coast. They are truly discretionary, as any West Coast or East Coast port that possesses the adequate facilities and services to satisfy a shipper's needs could receive this cargo. For the Eastern US, intermodal traffic has been developing, but not with comparable volumes to cargo moving off the West Coast into the east. Today, there are reverse land bridge flows, with some speculation about the magnitude of intermodal diversion from the West Coast to the Eastern United States.

In addition to rail corridors, several short sea shipping or container on barge operations have been explored in the United States. This includes the recent 64 Express, which operates a barge service between Richmond, Virginia and the Hampton Roads area. The failed New York to Albany Port inland water service represents an example of misaligning an intermodal corridor project without securing committed partners.

---

1   Author for correspondence.

Given the lessons learned, the US Government is currently exploring the importance of examining and improving operations along freight corridors for the next bill authorising the nation's highway transportation. There are discussions about examining multimodal corridors to manage highway traffic, emissions, and related externalities associated with commercial freight movement, but these research efforts are still in their infancy.

Major infrastructure projects represent long-term commitments and they have far-reaching implications for future transport operations. Legacy obligations therefore exert perhaps the most significant single influence on transport planning. There appears to be a real desire on the part of the US Department of Transportation to deliver a unified transport vision but due to other financial and statutory obligations the ability to deliver this vision may be compromised.

11.2   US Ports and Shipping

The dominance of the Los Angeles/Long Beach port complex can clearly be seen, although some shippers have diversified their gateways into the US in the aftermath of the 2002 West Coast labour lockout. Three new opportunities are arising that have the potential to influence port competition in the United States.

The first is the expansion of the Panama Canal, due to accommodate 13,000 TEU vessels by 2014. This will allow large vessels coming from the Far East to bring cargo for the eastern United States through the canal and directly into east coast or gulf ports (draft permitting). The port of Virginia at Hampton Roads is expecting to be the major beneficiary of this development. New York/New Jersey also has the requisite draft but is currently limited by air draft restrictions (although there is some talk of altering the offending bridge). Other ports in the Gulf and the Atlantic Seaboard are also struggling to get the necessary depth to receive these larger vessels. However the additional time taken to traverse the canal and reach the east coast may be unattractive to shipping lines. For example, to reach Chicago via Los Angeles/ Long Beach takes 14 days at sea plus five days on rail, whereas it takes approximately 25 days to reach Norfolk by sea from Shanghai, with an additional two days to Chicago.

Moreover, the role of the Los Angeles area as the largest manufacturing area in the country means that many forwarders will not want to forego the economies of scale that can be gained by transporting all their US cargo to this location then separating freight for inland destinations at this point for onward transportation by rail.

Secondly, the advent of the port of Prince Rupert in Canada provides a one-day shorter west coast option to shipping lines seeking to access North American markets. The port currently has a capacity of 500,000 TEU (but with room for expansion up to 2m TEU), and with sufficient depth to accommodate container ships up to 12,000 TEU (Fan et al. 2009). In 2009 the port handled 265,258 TEU (Containerisation International).

A third and (less important) development is the gradual westward movement of some manufacturing in the far east (to India, Thailand, etc.), leading to a potential scenario whereby the Suez Canal route to eastern American markets becomes time-competitive with the Pacific route to the west coast, and would require a shorter rail journey once the cargo is landed, not to mention removing the requirement to change from western to eastern railroads at Chicago.

All three developments will challenge the dominance of the San Pedro Bay ports, although it is unlikely that any of these changes have the potential to capture more than a small percentage of their cargo. However, as will be discussed below, hinterland access strategies of these ports will have a determinative impact on port competition.

Table 11.1   Top ten US container ports in 2009

USA Ranking	World Ranking	Port Name	Trade Region	Total TEU
1	16	Los Angeles	West Coast	6,748,994
2	18	Long Beach	West Coast	5,067,597
3	21	New York/New Jersey	East Coast	4,561,831
4	42	Savannah	East Coast	2,356,574
5	51	Oakland	West Coast	2,051,442
6	59	Houston	Gulf Coast	1,797,198
7	60	Virginia	East Coast	1,745,228
8	62	Seattle	West Coast	1,584,596
9	65	Tacoma	West Coast	1,545,855
10	102	Miami	East Coast	807,069

Source: Containerisation International

11.3   Rail Freight in the USA

There are three classes of railroads in North America: Class I (national), II (regional) and III (shortline). Not including passenger railroads (Amtrak in US and Via Rail in Canada), there are currently nine class I railroads (annual revenues in 2008 of over US$401.4 million) in North America. Seven operate in the USA: the big four (BNSF and UP in the west, CSX and NS in the east) plus the two Canadians (CN and CP) and the smaller KCS.[2] There are also two in Mexico: Ferromex and Kansas City Southern de México (wholly owned by Kansas City Southern). There is also a fourth class of railroad that performs switching and terminal operations.

---

2   For ease of reference, the following abbreviations are used. BNSF: Burlington Northern Santa Fe, UP: Union Pacific, NS: Norfolk Southern, CN: Canadian National, CP: Canadian Pacific, KCS: Kansas City Southern. Note that CSX is the full name and not an abbreviation.

2008 revenues at each of the Class I Railroads are shown in Figure 11.1. The difference in revenue between western (BNSF and UP) and eastern railroads (NS and CSX) is striking, arising from larger volumes greater distances with fewer interchanges.

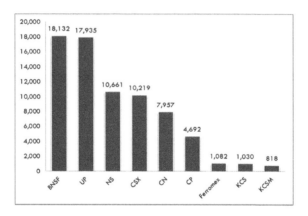

Figure 11.1   Annual revenues at class I railroads in 2008
Source: AAR, 2010a

Based on the Freight Analysis Framework, a national database integration effort, total freight transportation in the US is forecast to increase over the next thirty years. Over that same period of time, total international trade shipments will grow from 11 per cent of the weight of all shipments to 16 per cent of all shipments in 2040. For the railroads, international cargo as a share of total transportation by weight will increase fourfold (from 3 per cent to 12 per cent) (FHWA 2010a). Table 12.2 presents this data.

However, based on studies commissioned by the American Association of Railroads and the American Association of State Highway Transportation Officials, there are lingering questions about the future reliability of both the highway and rail networks in the future.

A number of key points need to be understood by non-US readers in order to gain an appreciation of the US system, addressed in turn below.

11.3.1   Economies of Scale Gained by Long, Double-Stacked Trains

The reason that rail has a higher market share in the US than in Europe is because it is the natural mode for long distance hauls, able to generate economies of scale. This is particularly the case for the western railroads, which enjoy longer distances and fewer interchanges than the eastern operators. Similarly, double-stacked capacity on many lines, in addition to train lengths of over 10,000ft in some cases mean that US trains can reach capacities of 650 TEU (compared to around 80–90

Table 11.2 Weight of shipments by transportation mode, 2007 and 2040 (million tonnes)

	2007				2040			
	Total	Domestic	Exports[b]	Imports[b]	Total	Domestic	Exports[b]	Imports[b]
Total	18,581	16,576	656	1,349	27,104	22,772	1,811	2,521
Truck	12,766	12,580	95	91	18,445	17,963	274	208
Rail	1,894	1,745	61	87	2,408	2,109	155	144
Other Modes	3,921	2,250	499	1,172	6,251	2,701	1,382	2,169
Water	794	360	52	382	1,143	482	105	556
Air, air and truck	13	3	4	6	41	5	16	19
Multiple modes and mail[a]	1,531	519	409	603	3,119	724	1,179	1,216
Pipeline	1,270	1,100	4	166	1,509	1,158	9	342
Other and unknown	313	269	29	15	440	331	73	35

[a] *Multiple* modes and mail includes export and import shipments that move domestically by a different mode than the mode used between the port and foreign location.

[b] *Data* do not include imports and exports that pass through the United States from a foreign origin to a foreign destination by any mode.

Source: FHWA 2010b

TEU in Europe). Therefore Class I railroads are profitable businesses, unused to government intervention, as they operate as private, and not public companies (also, American freight trains tend to focus on freight, and not passenger, movements, leading to a greater focus on network efficiencies, and not necessarily local transit movements).

11.3.2   Vertical Integration and Total Separation of East and West Markets.

In the US, railroads are vertically integrated, meaning that each company owns its own tracks and rolling stock and in most cases terminals. Therefore they operate completely separately from one another, although railroads may allow for track usage in certain situations. In Europe rail operating companies compete with each other on common-user track, forcing the maintenance issues into the public, and not private, sector. Furthermore, in the US, the east and the west of the country are entirely separate. BNSF and UP compete from the west coast to Chicago, while NS and CSX compete between Chicago and the east coast. The two Canadian railroads, CN and CP operate predominantly in Canada, although CN runs down from Chicago, through Memphis to the gulf of Mexico (more on CN in a later section).[3]

11.3.3   Role of Chicago as the Central Rail Hub

Due to historical development reasons, all six class I railroads mentioned above meet at Chicago. The city has some of the largest intermodal terminals in the world, each handling many hundred thousand lifts per year. The Chicago area includes approximately 900 miles of track and 25 intermodal yards, accommodating roughly 1,300 trains daily (CSX 2010). Approximately 14m TEU transited the metropolitan area in 2004 (Rodrigue 2008). This amount of traffic brings its own problems, but what is under consideration here is that freight needing to cross Chicago needs to be transported between east and west coast railroad terminals, either by rail or road. Whereas full trains that do not require reworking will change crew and power at the arriving terminal and then depart, trains carrying containers for more than one destination will need to be split and reassembled into new trains that may then need to be transported to another railroad across town. This reworking can take up to 48 hours; therefore "rubber tyre transfers" are more common. According to Rodrigue (2008), "about 4,000 cross-town transfers are made between rail yards each day averaging 40km each."

---

3   The University of Memphis has an interactive railroad map on their website, showing complete networks of each Class I railroad separately or together. https://umdrive. memphis.edu/haklim/public/final_the3rd.swf.

As cross town transfers grew more frequent, road congestion at grade crossings grew more severe.[4] In the winter of 1999/2000, a large snowstorm caused so much chaos for rail freight that a political tipping point was reached when the city exerted pressure to get the railroads to work towards a collective solution. This was the beginning of the CREATE (Chicago Region Environmental and Transportation Efficiency Program) project, which took shape over the next couple of years. CREATE is a public private partnership involving the federal DOT, the state of Illinois, the city of Chicago, all the Class I railroads (except KCS) and the passenger lines Amtrak and Metra. The project is an umbrella group, formed in order to seek funding for a number of individual engineering works, including: "six grade separations between passenger and freight railroads to eliminate train interference and associated delay; it includes 25 grade separations of highway-rail crossings to reduce motorist delay, and improve safety by eliminating the potential of crossing crashes; and it includes additional rail connections, crossovers, added trackage, and other improvements to expedite passenger and freight train movements nationwide." (FRA website) The estimated cost for the entire project is US$1.534 billion, US$232m of which will come from the railroads, described as "an amount which reflects the benefits (as determined by the Participating Railroads and agreed to by CDOT [Chicago Department of Transportation] and IDOT [Illinois Department of Transportation] prior to the execution of this Joint Statement) they are expected to receive from the Project." (CREATE 2005) The remainder of the funds are expected to be sourced through a variety of federal, state and local sources. $100m was received through the TIGER grant scheme (see below).

The operational issues discussed above have also led to some innovative responses by individual railroads. Both eastern railroads NS and CSX are developing hubs in Ohio (see below). Likewise, western railroads will sometimes rework trains before they reach Chicago, for example BNSF will sometimes do this at Clovis (New Mexico) or Fort Madden (Iowa). This strategy allows the train to go straight through to the eastern railroad terminals without being reworked at BNSF's Chicago yard. Another strategy has been pursued by CN, which purchased the old EJ and E line that bypasses the city down the west side. They began operating on this line in January 2009 and also sell paths on that line to other railroads. Therefore the future could see CN moving containers between Prince Rupert and their newly redeveloped hub at Memphis,[5] using the EJ and E line to bypass the congestion at Chicago.

---

4   There had been times when long trains being assembled on the mainline were blocking traffic at grade crossings for so long that police were putting parking tickets on the trains.

5   Memphis is also a key rail hub in the US, where five Class I railroads meet.

11.3.4   Role of International vs. Domestic Containers

The scale of domestic cargo in the US needs to be mentioned. As well as the millions of international TEU, 89 per cent of cargo in the US is domestic cargo (FHWA 2010a). Therefore this market dominates, and domestic cargo moves in 53ft boxes (as opposed to 40ft and 20ft maritime boxes). Therefore it makes sense both operationally as well as financially (i.e. it is cheaper per tonne for trucks and trains because fewer boxes are moved) to transload foreign cargo at or near the port from 40ft boxes into domestic 53ft boxes. (This also makes it expensive to reposition empty containers for outbound shipments from many hinterland markets that do not have sufficient inbound international container traffic.) About 25 per cent of all international cargo moved by rail is transloaded into these domestic containers (Rodrigue and Notteboom 2009). Therefore in the area surround Los Angeles/ Long Beach, millions of square feet of warehousing are located for these transloading activities. Additional reasons to transload include the fact that since the US is a net importer, taking a maritime container thousands of miles inland without an export load to send back means that the container will need to be shipped back empty to the port. Interestingly, Notteboom and Rodrigue (2009) speculate that due to the concentration of trade lanes between China and the USA, and since most of the 53ft containers are manufactured in China, perhaps in a decade there may arise the possibility of the 53ft maritime container. But in the meantime, existing investments in current ships designed for multiples of 20ft will inhibit that possibility.

11.3.5   Operational Differences

Inland terminals in the USA tend to be larger than in Europe, as railroads are used more as landbridges across the country than a network of small linked terminals like in Europe (Rodrigue and Notteboom 2010). A number of interesting operational differences may also be observed between Europe and the USA. In Europe, intermodal terminals are generally grounded facilities, meaning that containers are transferred between train and truck, and if a direct transhipment is not made, the containers are stacked on the ground. The truck driver will arrive at the terminal with his own chassis and the container will be lifted onto this. By contrast, in the US, both chassis and containers are owned by the carrier (be that the shipping line or 3PL), while the truck driver simply arrives in his tractor. Containers are loaded onto waiting chassis and the arriving driver will hook up to a loaded chassis and take it away. These wheeled facilities require a great deal more room as there is less equipment that can be stacked, but they can be quicker for the incoming drivers who do not have to wait for their container to be located in a stack. This also means that cranes make fewer unproductive moves to pick through a stack of containers

Yet the problem of having enough available chassis of the correct company has caused problems for terminal operators, and despite recent establishment of chassis pools (a number of owners sharing each other's equipment), there is now a move by some terminal operators towards the European model of grounded facilities. These require more use of cranes but utilise a far smaller footprint. Therefore they not only avoid the chassis problem but also account for the fact that land is not as available, nor as cheap as it was in the early days of intermodal terminal construction.

New terminals of both wheeled and grounded type are being built, largely with automatic stacking equipment. The 185-acre BNSF intermodal yard at Memphis, TN, opened in 2010, is a grounded facility. It represents a US$200m development with eight wide-span cranes, five for operating the working tracks and three for the stack. The site has 48,000 feet of track with enough length to work a full train without cutting. Likewise, the new CSX terminal at North Baltimore (due for completion in 2011 as part of the National Gateway project – see below) will be a grounded facility of similar design. Both sites will have capacity to handle over 500,000 containers per year. By contrast, the NS intermodal terminal at Rickenbacker near Columbus, Ohio (opened in 2008 as part of the Heartland Corridor project – see below) is a wheeled facility covering 125 acres (with 175 acres for development) with a capacity of 400,000 containers annually and 40,580 feet of track.

Rationalisation of the rail business in the 1980s (see discussion on the Staggers Act, below) resulted in fewer, larger intermodal terminals. The rule of thumb for rail operators is now about a minimum of 100,000 lifts annually for a feasible terminal. Therefore shippers will need to locate near these large sites to gain access to the main trunk routes, as has occurred in the Rickenbacker International Airport development.

In terms of port operations, the large shipping lines will have their dedicated terminals (therefore not operated by a railroad). On-dock railheads are very expensive to run because the trains must be loaded using longshore labour, which also raises issues as to whether the trains are loaded to the exact specifications of the rail operator. Shipping lines may not have enough volume for one location to fill daily train loads direct from their terminal so often there is no genuine need for on-dock rail. Therefore in some cases it may be more effective to dray the container a couple of miles to a near-dock facility where the containers can be consolidated from multiple terminals into trains bound for each location. If a full load is possible then on-dock is more efficient and cheaper, therefore it can help to make a rail move more competitive than road (because it removes the pre-haul), but as this is not always the case, using on-dock rather than near-dock can be problematic.

11.3.6   Other Relevant Issues

The Rail Safety Improvement Act of 2008 required all railroads that carry passengers or toxic materials to implement Positive Train Control (a safety process that uses GPS to prevent train accidents by monitoring their movement and overriding their progress if it is deemed unsafe) by 2015. The legislation was controversial in the industry as the railroads are required to fund it themselves, at a cost of over US$5bn for installation alone, resulting in a total cost of around US$13.2bn for installation and maintenance over 20 years (AAR 2010b).

A second issue that may impact on the rail industry is the new safety criteria for truck drivers based on proposed changes to the current "Hours of Service" rules and other driver licensing requirements. The safety record will now follow each truck driver from state to state as well as between jobs. The likely result of this change is that insurance rates will probably go up and the driver workforce will potentially fall. These additional costs placed upon the road haulage industry are likely to make rail freight more attractive.

11.4   Government Policy and Legislation

In order to understand the current system of intermodal transport and the role of policy and regulation in the US, a brief overview of legislation to this point is required. Two key issues often raised by intermodal transport operators include the Jones Act (1920) which requires that any vessel operating between two US ports must be US-built, -owned and -manned, and the Harbour Maintenance Tax (HMT). Originally introduced in 1986, the HMT is a federal tax imposed on shippers based on the value of the goods shipped through ports. Its purpose is to fund maintenance and dredging of waterways, which are the responsibility of the US Army Corps of Engineers. Perakis and Denisis (2008) discuss the obstacle that HMT presents to the development of short sea shipping in the US. As the tax is applied at every port, a water leg in an intermodal chain will attract this fee whereas transloading to road or rail will not.

Although container operations were developing in the 1960s and 1970s, several laws passed in the early 1980s allowed for cooperation between different transportation groups to develop. One law was the Staggers Act of 1980, which partially deregulated some areas of the railroad industry. The Staggers Act reduced the number of crewpersons needed for each train, which lowered total labour costs for each train. Secondly, the Staggers Act removed several pricing and scheduling limitations, which increased the railroad's ability to become flexible in changing to market needs. These changes were designed to make the railroads more competitive for long distance domestic freight that had been lost to truck companies during the 1970s. As in other countries, this eventually led to a number of mergers and there are currently 4 major railroads in the US (see above). The Ocean Shipping Act of 1984 relaxed many restrictions faced by the carrier operators and allowed an ocean carrier to provide inland distribution on a single through bill of lading.

The Intermodal Surface Transportation Efficiency Act (ISTEA) (1991) heralded something of an intermodal approach to highway and transit funding, including collaborative planning requirements (Chatterjee and Lakshmanan 2008). It provided supplementary powers to metropolitan planning organisations and designated High Priority Corridors in the National Highway System. In 1998, the Transportation Equity Act for the 21st Century (TEA-21) authorised the federal transport programme until 2003. The act required a number of planning objectives for regional transport plans, including safety, economic competitiveness, environmental factors, integration and quality of life. However despite these attempts to foster an intermodal approach to transport planning, key government agencies (such as DOT departments) and industry bodies remain modally-based. (Holguin-Veras et al. 2008).

11.5   New Developments in Government Policy and Legislation

At this point, US federal freight policy is moving towards a more integrated transportation system, however, the necessary funding to do so remains in various agencies that do not necessary have the authority to work on cross-jurisdictional projects. In some cases, modal agencies may be more responsible for safety and not infrastructure investment, while in other cases the role of regulatory oversight may reside in different agencies, depending upon the area of concern. Finally, thin the US, the infrastructure for waterways is actually managed by the US Army Corps of Engineers, and not the US Department of Transportation.

Normally the federal DOT gives money to the state DOTs and they decide how to spend it, but this system creates little incentive for states to spend money on projects that are perceived to be of primary benefit to other states. Therefore there has been a realisation at national level that attention should be paid to cross-border projects. This led to the projects of national and regional significance (see below). This was quite a new development, and especially for railroads to be eligible rather than just road projects.

The Safe, Accountable, Flexible and Efficient Transportation Act: A Legacy for Users (SAFETEA-LU 2005) introduced approximately US$1.8bn in congressional earmarks, designated projects of national and regional significance. These were large infrastructure project funds decided by Congress. Benefits could include improving economic productivity, facilitating international trade, relieving congestion, and improving safety. One example is the Heartland Corridor, which will be discussed below.

The stimulus package, named the American Recovery and Reinvestment Act (2009), provided US$1.5bn for transport projects through the Transportation Investment Generating Economic Recovery (TIGER) programme. This money was available for all transportation projects (not just freight) and would be awarded on a competitive basis, with applications due in September 2009 and announcements made in February 2010. Private money in TIGER applications was matched by public money. The five major goals for TIGER grants were economic competitiveness, safety, state of good repair, liveability and environmental sustainability. This was the first time money was awarded in this fashion, and a second round of US$600m was awarded in September 2010. The popularity of the funding programme meant that the DOT was swamped with applications for each round of funding, receiving almost 1,500 applications totalling nearly US$60bn for the first round, and almost 1,000 applications totalling US$19bn for the second round. The recent experiences of the TIGER Program, where the Department of Transportation solicited many multimodal projects (both for passenger and freight) demonstrated the need for such programs, but also the lack of guidance available to engage in such broad comparisons.

The result of this round of funding has been a significant revival of interest in rail projects. As states were the only eligible applicants, Class I railroads were required to form partnerships with the states in order to process an application. The list of recipients indicates that taking an integrated approach to transport problems by focusing on corridors was considered an attractive proposition for federal legislators. The majority of awards were for public transit programmes, highways and other infrastructure upgrading, however freight-specific projects such as transportation hubs and port upgrades also received financial support. Some larger freight projects included US$98m for the CSX National Gateway project (see below), US$100m for the Chicago CREATE project and US$105m for the Norfolk Southern Crescent Corridor (see below). A marine highways project in California was also among the recipients.

The Energy Independence and Security Act (2007) included a provision for "America's Marine Highway Program" to integrate the nation's coastal and inland waterways into the surface transportation system. Therefore Congress instructed the DOT Maritime Administration (MarAd) to create a Marine Highway program that examined ways to utilise waterways where they may provide some services on parallel highway routes to alleviate bottlenecks. They assessed the country's waterways and invited applications and in August 2010 eventually designated 18 marine corridors, eight projects, and six initiatives for further development. These eight projects are now eligible to bid for a total of US$7m in funding which, while clearly not enough for an infrastructure project will contribute towards future service development and will receive preferential treatment for future federal funds.

As will be seen in the following discussion, there is an increasing focus on infrastructure corridors. As these projects encompass a number of localities, regions and states, not to mention private and public stakeholders, they have necessitated new methods of management and new funding schemes. However the real role for this money is to enable large consortia to come together where public and private benefits can be clearly identified amongst all the parties. In reality the federal government could never spend enough money to exert significant influence on the operations of the rail industry. Indeed, railroads in the past have been reluctant to accept public money for fear of strings being attached.

In addition, some problems will be solved by the market without the need for government intervention. The chassis problem is a good example. As discussed earlier, chassis management has become a problem over the last decade but the issue is beginning to resolve itself, first through chassis pools and then through the growing trend of grounded facilities. In this case, policy intervention has not been required as a combination of operational efficiency on behalf of railroad terminal operators and the increasing costs to carriers to maintain chassis fleets have resulted in new methods of operation.

Therefore it can be seen that different issues can best be solved by different approaches, whether through policy intervention, planning strategies, operational changes or market forces. The difficulty for politicians and planners lies in recognising which freight issues can best be solved by which measure.

11.6   Corridors: Current and New Developments

In order to improve hinterland access with the aim of capturing or retaining inland markets, a number of rail corridor projects have been undertaken with involvement of US ports to greater or lesser extents. The most significant of these will be discussed below.

11.6.1   Alameda Corridor

The Alameda Corridor Transportation Authority (ACTA) is a joint powers authority that was set up by the ports of Los Angeles and Long Beach in 1989. The ports purchased the required rail lines from the railroads and the crucial factor in the project was that the railroads agreed to use the corridor once it was built.

The Alameda Corridor is a short (20 miles) high capacity (3 double-stack tracks) line designed to reduce congestion and other negative externalities associated with the extremely high container flows of the San Pedro Bay ports (Los Angeles and Long Beach – combined 2009 throughput of 11.8m TEU). The project consolidated four branch lines, reduced conflicts at 200 grade crossings and included a ten-mile trench. The line was opened in 2002, with a capacity of about 150 trains per day.

The total cost of the project was US$2.43bn, split between US$1,160m revenue bonds, US$400m federal loan (the first of its kind), US$394m from the ports, US$347m MTA grants and US$130m from other sources. (ACTA 2010) The ports are directly involved in the project, as they are the financial guarantors of the corridor and will lose money if the route is not used and incurs losses. While the two ports are located immediately next to each other and have acted together in this instance, they remain separate institutions, each administered by their respective city's harbour department (Jacobs 2007).

In 2009, of the 11.8m TEU through the ports, 3.4m TEU travelled up the corridor (2.8m TEU using on-dock connections and 0.6m TEU near-dock). 0.7m TEU used off-dock rail, 3.4m TEU used rail after transloading into 53ft domestic containers and 4.3m TEU travelled inland by truck. (ACTA 2010)

Yet while the corridor solves certain problems for the port, it presents other issues for the two competing rail operators. UP operates a large intermodal yard (ICTF) in Carson – about 4–5 miles from the port, therefore they are able

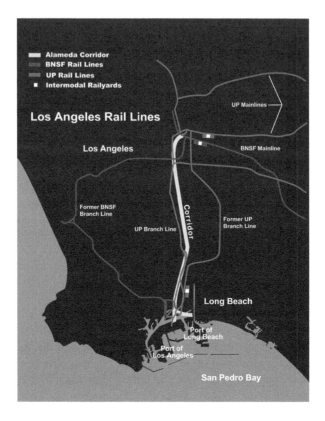

Figure 11.2   Map showing Alameda Corridor
Source: ACTA 2010. Graphic courtesy of the Alameda Corridor Transportation Authority

to truck their containers there for consolidation on block trains, then send those up the Alameda Corridor and across the country. However BNSF's main yard is at the end of the corridor, in Los Angeles. Therefore BNSF often drive trucks to transloading warehouses, then truck the 53ft container to their LA yard, thus bypassing the corridor. BNSF wants to build a new yard near UP's ICTF yard. This situation relates back to the early days of intermodalism, when SP (later bought by UP) bought the ICTF yard, as much to do with the location of the site relative to their business as clairvoyance with regard to the bright future of intermodal transportation. BNSF has small rail sidings near the port as well as an agreement with Maersk to use space within their LA port terminal, whereas UP has, in addition to their ICTF terminal, large rail sidings between the port and ICTF. Therefore when rail corridors are built, it is important to understand issues such as train marshalling that can have major impacts on usage of the mainline.

11.6.2   Alameda Corridor East

The Alameda Corridor East is primarily a highway programme, with the aim of removing a number of grade crossings (including a 2.2-mile trench that will result in the lowering of a 1.4-mile section of track) on the rail line heading inland from the Alameda Corridor terminus at Los Angeles. It will reduce congestion because, as rail traffic has priority (the rail tracks are on their own private land), road traffic has to wait whenever a train passes. A secondary aim is to increase safety through avoiding potential collisions. US$1.14bn has already been spent, with an estimated US$0.3–0.6bn required to complete the remaining grade separations. Of the money already spent, US$219m came from federal sources (including US$134m from a TEA-21 earmark and US$65m from SAFETEA-LU), US$536m from state sources, US$340m from the MTA (metropolitan transportation authority), US$23m from the city/county and US$20m from the railroad Union Pacific. (ACE 2010).

Callahan et al. (2010) criticised the project for not utilising the same institutional framework as ACTA, but realistically the facts of the situation precluded such a framework. Firstly, the grade separations do not actually provide much of an advantage for the railroad (as they already have the right of way), beyond the removal of the threat of collisions. Indeed, improving road flows could

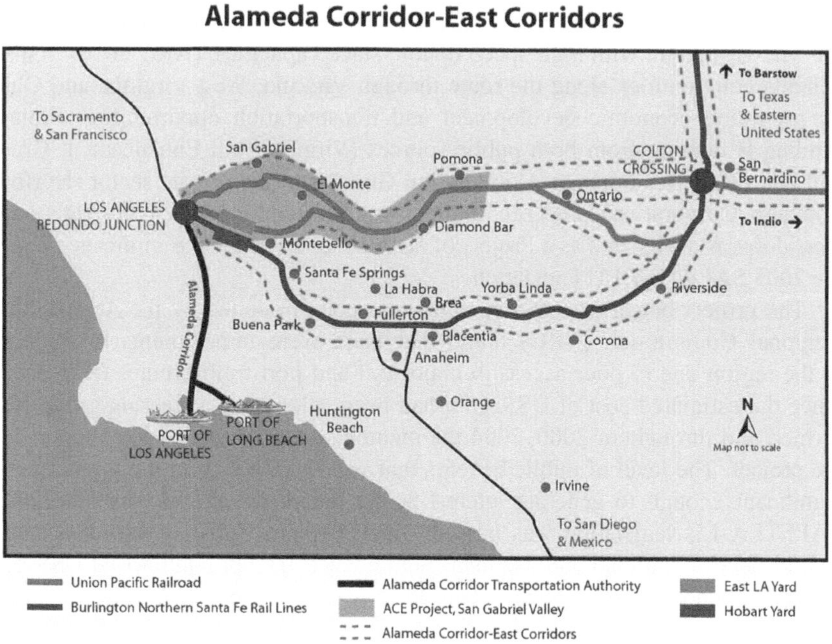

Alameda Corridor-East Corridors

Figure 11.3   Map showing the Alameda Corridor East project area
Source: ACE Construction Authority

be considered to provide an advantage to road haulage (Callahan et al. 2010). Furthermore, because the project does not consolidate freight on a new line, it was not possible to purchase the tracks under a new authority as in the case of ACTA. However in terms of the actual construction work, the project has been managed under a single joint powers authority ACE Construction Authority, which has been successful in utilising the branding of the corridor to focus public money that may have been more difficult to attract for each construction project individually. The project was unsuccessful in its application for TIGER I and II funding, although a grade crossing further inland at Colton was awarded US$33m through an application by Caltrans, the state transport authority.

11.6.3   Heartland Corridor

The Heartland Corridor, linking the Hampton Roads area to Columbus Ohio, and eventually to Chicago, represents the first multistate private–public rail corridor in the US. The work involves upgrading an existing coal line with restricted dimensions to handle international maritime and domestic double-stack container traffic moving from the Virginia Port Authority west through Virginia, West Virginia, and Ohio, continuing to Chicago and its interchanges with the western Class I railroads.

The project affords a significant competitive advantage to Virginia's ports by providing a shorter (by over 200 miles) and faster (by about 24 hours) route to the Midwest along with high-speed double-stack capacities. (ARC 2010). It also benefits communities along the route through Virginia, West Virginia, and Ohio by providing economic development and transportation opportunities. Project funding is coming from both public sources (Virginia Rail Enhancement Grant and Ohio Rail Development Commission Grant) and the private sector (Norfolk Southern Corporation). Federal funding was forthcoming when the Heartland Corridor was designated as a Project of National and Regional Significance under the 2005 SAFETEA-LU legislation.

The project began in 1999 when a study commissioned by the Appalachian Regional Commission (ARC) found that there were impediments to shippers in the region due to poor access to major rail and port traffic routes (RTI 2000) Once the estimated cost of US$200m had been calculated, a steering group was formed and throughout 2000–2004 the members worked to develop support for the project. The level of public benefits that would accrue from the project were significant enough to generate interest at the federal level and when the 2005 SAFETEA-LU legislation was passed, the Heartland Corridor was designated as a Project of National and Regional Significance. The bill authorised US$95m in federal funds for the project (reduced to US$84.4m by estimated obligation limitations, rescissions, etc.). Of the total cost of US$195.2m, US$84.4m was federally funded, US$101.0 million was contributed by NS, US$0.8m from the Ohio Rail Development Commission (ORDC) and US$9.0m came from the Virginia Department of Rail and Public Transportation (VDRPT).

Two Memoranda of Agreement, one between Federal Highway Administration (FHWA) Eastern Federal Lands Highway Division and Norfolk Southern, and the other between FHWA, EFLHD and the three states were completed in August 2006. The agreements identified roles and responsibilities for the Environmental Planning, Design and Construction of the Heartland Intermodal Corridor Project. Furthermore, the MoA established an unprecedented funding mechanism between the federal government and the railroads that allowed money to flow directly to the railroads from the federal government. A key aspect of the MoA was that since the majority of the tunnels were in West Virginia, the other states had to agree that the majority of the money would be spent there, as they would all benefit. It was also agreed that federal money would only be used on the track work. Intermodal terminals would need to be funded from other sources. In addition, in order to access the APM terminal in the port of Norfolk, a separate US$60m project was required to relocate the Commonwealth Railway short line.

Along the route, 28 tunnels and 26 other overhead obstructions needed to be raised to allow the passage of double-stacked container trains. Construction began in October 2007 and the first double-stack train ran on 9 September 2010. A large intermodal terminal was opened at Rickenbacker in 2008 (developed through a US$68.5m partnership between NS and the Columbus Regional Airport Authority, US$28m coming from the US DOT) and sites are being developed at Prichard WV and Roanoake VA.

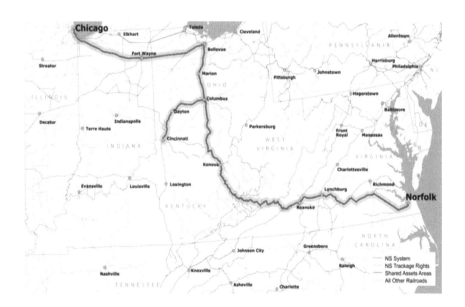

Figure 11.4   Map showing the Heartland Corridor route
Source: Norfolk Southern website

11.6.4 National Gateway

The National Gateway project bears certain similarities with Norfolk Southern's Heartland Corridor. Both will connect the Virginia Port Authority to the state of Ohio, and then to Chicago, and both will be anchored around a new large terminal in Ohio, both of which were discussed in an earlier section. Both corridors involve a number of clearance upgrades to allow double-stack capability.

The National Gateway[6] is a Public Private Partnership (PPP) that involves 61 double-stack clearances, the construction or expansion of six intermodal terminals and will cost US$842m (CSX 2010), including US$98m in funding from the first TIGER programme.

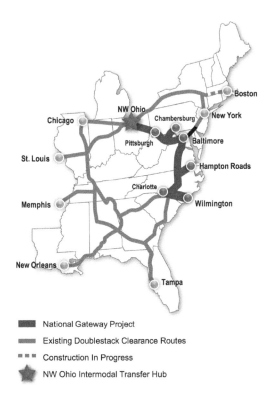

Figure 11.5   Map of CSX National Gateway route
Source: CSX 2010

---

6   For more information visit http://www.nationalgateway.org/.

CSX is also attempting to improve connectivity of landbridge services by increasing partnership with western railroads UP and BNSF, aiming to provide a seamless door-to-door service for customers. They are talking to western railroads to run a dedicated train of CSX cars past Chicago to their north Baltimore hub and they will then split it there, rather than having to marshal the train in the Chicago terminal. This kind of relationship represents a new development for the railroad and is perhaps indicative of greater cooperation between operators.

11.6.5   Other Corridor Projects

More recently, the rubric of intermodal corridors is being used to market various upgrades to the US rail network. NS have proposed a number of projects along what they call the Crescent Corridor, a 1,400-mile stretch running between New Orleans and New York. They are aiming to develop a PPP to cover the estimated cost of US$2.5bn. The project involves 13 states, 11 new or expanded terminals and 300 miles of new track. In February 2010 the project was awarded US$105m in TIGER I grants through an application from Pennsylvania DOT and in August 2010 six states submitted applications under the TIGER II programme totalling US$109.2m, although none were successful.[7] Another corridor project is the Meridian Speedway, a joint venture between KCS and NS over a 320-mile stretch of track upgrade costing US$300m, agreed in 2006.[8]

11.7   Conclusion

Intermodal freight transport is developing in the United States, particularly through the development of a corridor approach (Rodrigue 2004). This integrated approach brings together public and private actors in order to direct investment to infrastructure improvements, while aiming to retain the benefits of private sector operation. While public money is targeted at infrastructure development, a crucial aspect of any corridor project is the alignment of terminal facilities with local and regional demand, with a particular focus on the co-location of transport facilities with logistics parks (Rodrigue et al. 2010).

Vital for the successful development of such multi-partner, or indeed multi-region or multi-state projects is the agreement amongst stakeholders that the investment will benefit all locations along the corridor (McCalla 2009). Thus information sharing and relationship building, along with promotion and branding activities are essential. In addition, innovative planning and funding mechanisms may be required, and the most positive indication that US freight capacity issues are being addressed is the recent development of such mechanisms.

---

7   For more information visit the website. http://www.thefutureneedsus.com/crescent-corridor/.

8   http://www.kcsouthern.com/en-us/Media/Pages/MeridianSpeedway.aspx.

However ongoing work will be required to continue to negotiate these mechanisms at local, state, regional and national level. Moreover, as such large corridor projects can easily take a decade or more to come to fruition, planners at all levels require a detailed understanding of the freight needs of their area in order to initiate the necessary infrastructure developments for the next thirty years. Given that current planners also have to deal with the legacy obligations resulting from the decisions of previous planners, as well as juggle policy directives from politicians and operational requirements from private operators, the task remains formidable.

11.8   Acknowledgements

The authors would like to thank all the interviewees who shared their time and knowledge with us over the course of numerous interviews conducted in September 2010.

Research in the USA was conducted with financial support from the Royal Society of Edinburgh and the Interreg IVb North Sea Region-funded Dryport project.

11.9   References

AAR 2010a. *Class I Railroad Statistics* [online]. Available at: http://www.aar. org/~/media/aar/Industry%20Info/AAR%20Stats%202010%200524.ash [accessed 26 October 2010].

AAR 2010b. *Positive Train Control* [online]. Available at: http://www.aar.org/ Safety/~/media/aar/backgroundpapers/positivetraincontrol.ashx [accessed 8 November 2010].

ACE 2010. *Alameda Corridor East*, San Gabriel Valley. Irwindale: ACE.

ACE website [online]. Available at: http://www.theaceproject.org/photo/ intermodel1.jpg [accessed 3 November 2010].

ACTA 2010. Presentation given by Art Goodwin, October 2010.

ARC 2010. *The Heartland Corridor: Opening New Access to Global Opportunity.* Washington, DC: ARC.

Callahan, R.F., Pisano, M. and Linder, A. 2010. Leadership and Strategy: A Comparison of the Outcomes and Institutional Designs of the Alameda Corridor and the Alameda Corridor East Projects. *Public Works Management and Policy*, 14(3), 263–287.

Chatterjee, L. and Lakshmanan, T.R. 2008. Intermodal Freight Transport in the United States, in *The Future of Intermodal Freight Transport,* edited by Konings, R., Priemus, H. and Nijkamp, P. Cheltenham: Edward Elgar, 34–57

Containerisation International 2010 [online]. Available at: http://www.ci-online. co.uk/default.asp [accessed 8 November 2010].

CREATE 2005. *CREATE Final Feasibility Plan.* Chicago: CREATE.

CSX 2010. Presentation given by Parker McCrary, September 2010.

Fan, L., Wilson, W.W. and Tolliver, D. 2009. Logistical Rivalries and Port Competition for Container Flows to US Markets: Impacts of Changes in Canada's Logistics System and Expansion of the Panama Canal. *Maritime Economics and Logistics*, 11(4), 327–357.

FHWA 2010a. *Freight Facts and Figures 2010* [online]. Available at: http://ops. fhwa.dot.gov/freight/freight_analysis/nat_freight_stats/docs/10factsfigures/ pdfs/fff2010_highres.pdf [accessed 2 February 2011].

FHWA 2010b. *Freight Analysis Framework*, version 3.1, 2010. US Department of Transportation, Federal Highway Administration, Office of Freight Management and Operations.

FRA website [online]. Available at: http://www.fra.dot.gov/rpd/freight/1486.shtml [accessed 4 November 2010].

Holguin-Veras, J., Paaswell, R. and Perl, A. 2008. The Role of Government in Fostering Intermodal Transport Innovations: Perceived Lessons and Obstacles in the United States, in *The Future of Intermodal Freight Transport,* edited by Konings, R., Priemus, H., Nijkamp, P. Edward Elgar: Cheltenham.

Jacobs, W. 2007. Port Competition between Los Angeles and Long Beach: An Institutional Analysis. *Tijdschrift voor Economische en Sociale Geografie.* 98(3), 360–372.

McCalla, R.J. 2009. Gateways are More than Ports: The Canadian Example of Cooperation Among Stakeholders, in *Ports in Proximity; Competition and Coordination among Adjacent Seaports,* edited by Notteboom, T.E., Ducruet, C., de Langen, P.W. Farnham: Ashgate.

Norfolk Southern website [online]. Available at: http://www.thefutureneedsus. com/site/download/?hr=/images/uploads/Heartland_Corridor_Map_hi-res.jpg [accessed 26 October 2010].

Notteboom, T., Rodrigue, J.-P. 2009b. The Future of Containerization: Perspectives from Maritime and Inland Freight Distribution. *GeoJournal.* 74(1), 7–22.

Perakis, A.N., Denisis, A. 2008. A Survey of Short Sea Shipping and its Prospects in the USA. *Maritime Policy and Management.* 35(6), 591–614.

Rodrigue, J.-P. 2004. Freight, Gateways and Mega-urban Regions: the Logistical Integration of the Bostwash Corridor. *Tijdschrift voor Economische en Sociale Geografie.* 95(2), 147–161.

Rodrigue, J.-P. 2008. The Thruport Concept and Transmodal Rail Freight Distribution in North America. *Journal of Transport Geography.* 16(4), 233–246.

Rodrigue, J.-P., Debrie, J., Fremont, A., Gouvernal, E. 2010. Functions and Actors of Inland Ports: European and North American dynamics. *Journal of Transport Geography.* 18(4), 519–529.

Rodrigue, J.-P., Notteboom, T. 2009. The Terminalisation of Supply Chains: Reassessing the Role of Terminals in Port/Hinterland Logistical Relationships. *Maritime Policy and Management.* 36(2), 165–183.

Rodrigue, J.-P., Notteboom, T. 2010. Comparative North American and European Gateway Logistics: The Regionalism of Freight Distribution. *Journal of Transport Geography.* 18(4), 497–507.

RTI 2000. *Transportation and the Potential for Intermodal Efficiency Enhancements in Western West Virginia.* Report prepared on behalf of the Appalachian Regional Commission, the West Virginia DOT and West Virginia Planning and Regional Development Council. Huntington: RTI.

Chapter 12

Implementing Dedicated Areas for Foreign Trade in the Santos Metropolitan Region: The Brazilian Experience

Leo Tadeus Robles

12.1   Introduction

Foreign trade business has always been important to the Santos Metropolitan Region (RMBS) where Santos Port, the main Latin American Port,[1] is located and through which around 35 per cent of Brazilian foreign trade flow (in value) passed by. After the deregulation of Brazilian port activities (Law no. 8,630 from 26 February 1993 – Brazilian Port Modernization Law) deep institutional and technological changes have occurred in the port business. Containers have facilitated goods handling, and information and communication technologies have sped up relationships and documentation flows. Customs issues have seen an improvement and have moved towards more efficient logistics management.

This technological innovation, has caused, as in other ports around the world, direct job reduction and transformed port works, demanding from port workers the ability to deal with sophisticated and expensive equipment. Another related issue is the customs information systems that control imported or exported goods flows. In this sense, the opportunity for locating areas and facilities dedicated to foreign trade linked activities has presented itself to port cities or regions, specifically via the so called Special Customs Regimes Areas, which have the potential to enhance steady job creation in port cities. In Brazil, the workforce in port cities is seasonal, due to summer vacations.

This chapter analyses the Brazilian experience of implementing dedicated areas for foreign trade with special customs regimes. It is considered as a basic fact that Special Customs Regimes are a part of a country's public policies aimed at improving conditions which then enable it to better compete in international commerce. In specific areas they can also create new jobs and improve regional development.

It is in this sense that the Santos Metropolitan Region – RMBS is analysed: its potential to implement special customs regime areas for goods handling, as well as its preparation for foreign trade and logistics service provision. Recent Brazilian

---

1   Santos Port is the main Brazilian Port in terms of goods movement in value and, in Latin America, in terms of container (TEUs) movements (CODESP, 2010).

initiatives for implementing areas dedicated to foreign trade are Processing Exportation Zones – ZPEs, dry ports, bonded warehouses, and Customs Industries and Logistics Centres – CLIAs. The difference between these initiatives is basically down to the specific legislation applied, as will be discussed later in this chapter, but the main objective of each is to provide incentives and facilitate foreign trade.

RMBS is located in São Paulo State (Figure 12.1), in the Southeast Region, the main Brazilian economic area, which with 42 per cent of the country's population accounts for almost 59 per cent of GNP. Santos Port, the main Latin American port, is a multipurpose port and, in 2008, the total cargo moved was 81.1mt, including 2,674,975 TEUs. (CODESP 2010) Containers movement increased by 126 per cent from 2001 to 2009, reaching 2,252,188 TEUs. This remarkable increase was due to private investment and Brazil's foreign trade development.

In 2009, according to the CODESP 2010 report, the total cargo moved in tonnes increased 2.5 per cent compared with the previous year, but container movement fell by 15.8 per cent. This decrease is related to the impact of the world financial crisis, which affected Brazilian value added exports and imports. However, in Santos Port's total movement figures, the decrease was compensated by the export of commodities such as soybean and sugarcane.

This study can be considered as exploratory. It is based on a literature review and analysis of academic and legal documents and specialised sites. In addition, sector executives were interviewed in order to identify projects and to discover their opinion as to what are the main obstacles to an effective implementation of such specialised areas in Brazil and, particularly, in RMBS. The qualitative approach is due to the fact that regulation is currently undergoing change which is being led by the Brazilian Congress and Government, and this change has generated some controversy among interested parties, as this chapter intends to show. The RMBS projects were identified in the interviews with city representatives and private companies involved with Special Customs Regimes initiatives.

Figure 12.1   Brazil: Santos Port location
Source: Author.

12.2   Brazilian Special Customs Regimes

The Secretariat of Foreign Trade (SECEX) in the Ministry of Development, Industry and Foreign Trade is in charge of proposing foreign trade policies and programmes, including import measures and procedures, to the Chamber of Foreign Trade (CAMEX), which has ultimately been responsible for foreign trade issues. The SECEX is also responsible for elaborating the legal terms required for the implementation of import measures. The Ministry of Finance, through the Secretariat of Federal Revenue (SRF), is responsible for customs administration (WTO 2004, p. 38).

Brazilian Customs regulation is divided into three categories: "common", "special" and "special applied to physical areas or dedicated sites". Common refers to duty payment without any particular proceeding. The special category allows goods import and export, with payment exemption or suspension. The last category allows goods transfer to physical areas, known as secondary zones (Faro and Faro 2007).[2]

Special Customs Regimes in Brazil refer to incentive mechanisms to enhance foreign trade by providing federal, state or local tax suspension. The main categories are drawback, temporary admission, customs transit, bonded warehouses, Special Customs Areas (Special System of Industrial Depots subject to Standardised Control – RECOF; areas for goods to be exported – REDEX) and special regimes for the importing of port equipment – REPORTO.

These regimes aim to incentive foreign trade and operate on the principle of not levying import or export taxes. For example, the Brazilian drawback system provides for the suspension, exemption or restitution of import and other taxes, when the imported goods, inputs or parts are used to produce exportable goods or to package them. The beneficiaries of these systems are industrial or commercial enterprises engaging in foreign trade.

Temporary admission allows the country permanence for a certain time period to goods coming from abroad for an established purpose without import tax payment. Customs transit allows merchandise transit, under customs control, from one port to another in the customs territory, with tax payment suspension. Bonded warehouse regimes allow, during import or export process, the merchandise to be deposited in authorised sites, with tax payment suspension. RECOF suspends the payment of import taxes and tax on merchandise imported for the purposes of industrialisation and goods production for export.

These regimes, as well as the previously mentioned Special Regimes, are strongly based on Information Technology (IT) facilities, including equipment, communication devices and new dedicated software.

---

2   Primary zone: ports and airports customs authorized, as well frontier customs points. Secondary Zone: the remaining customs territory.

As it is known, IT advances and cost reductions have been remarkable in the last decades, and their suitable application to customs procedures is a consensus as, in short, they represent information exchange among agents involved in foreign trade: importers, exporters, shippers, customs agencies, transporters, and so on. Foreign trade can be considered the result of three basic physical and financial flows based on information exchanges among the agents involved.

This chapter analyses the Brazilian conditions that led to the implementation of Special Customs Regimes in dedicated areas, especially, dry ports, customs industries and logistics centres – CLIAs, Export-Processing Zones – ZPEs, the recent Brazilian government studies initiative, the Logistics Activities Zones – ZAL, and other foreign trade facilitating initiatives such as the Authorised Economic Operator intended implementation.

12.3   Special Regimes Applied to Dedicated Areas

In Brazil, dedicated areas with special customs regimes are known by different names with differences in the beneficiaries involved and in the associated implementation legislation. SRF´s Normative Instruction no. 241/2002 indicated the sites allowed to implement customs industries to be dry ports and customs areas located at airports or ports (public or private terminals under the control of a Port Authority), with all of them required to have SRF authorisation. The main issue is that SRF determines the site location and service provision characteristics when authorising dedicated areas with customs special regimes.

The main area with Special Customs Regimes is Manaus Free Trade Zone (ZFM), located in the North Region (Amazon State), established in 1967 and presently comprising a manufacturing park, a trading centre and an agricultural and ranching district. The MZF is managed and promoted by the Manaus Free Trade Zone Superintendence (SUFRAMA), a government agency under the Ministry of Development, Industry and Foreign Trade. Companies located in ZFM may benefit from tax exemption for products to be consumed and/or manufactured. This is a consolidated experiment and it will not be analysed in this chapter, which focuses on other Brazilian alternatives to implement these kind of projects.

The Bonded Warehouse Regime, also known as a Dry Port, refers to an authorised area in the hinterland as well as a Bonded Warehouse located at sites of important industrial firms. This kind of special regime has existed since the 1970s when it was created to speed up customs processes and to reduce port congestion. They could be located on secondary zones as well primary zones. (Ricupero and Oliveira 2007).

In Brazil, there are 67 areas, 27 of them located in São Paulo State and four in RMBS. Typically, imports represent 70 per cent of a dry ports' movement and they are responsible for 20,000 direct jobs and four million TEU movements, comprising a merchandise value of around US$70 billion/year. (Cardoso 2007). These are private sites dedicated to moving, storing, and provide other services and operations related to customs procedures under SRF control. They are not allowed to occur in a primary zone. From 1971 to 1995, these services were authorised by SRF, but since July 1995 (Law 9,074), dry ports implementation has been submitted to public concession biddings and been considered as a public service. (Lourenço 2008)

In the last years, Brazilian legislation regarding Dry Port entities is evolving, but slowly and with difficulty. This fact could be linked to different interests prevailing in the sector, in spite of a consensus that these entities could facilitate and enhance foreign trade and could be important for regional or local economic development. Thus, these customs dedicated areas are receiving different names (other than dry ports) and are incorporating logistics services, as will be shown in the next section.

12.4   Customs Industries and Logistics Centres (CLIAs)

The CLIAs were created by the Federal Government by the Provisional Measure no. 320[3] on 24 August 2006 (PM 320) in order to restructure the port customs and logistics sector and the dry ports' legal and institutional model. (Lourenço 2008). The CLIAs regulation allowed aggregating value to merchandise by slight modifications and/or assembling operations, besides the activities of movement, storage, products processing, customs activities and tax payments.

This Special Customs Regime dedicated to physical areas was intended to permit customs industrial operations, such as production lines being installed at customs sites and having permission to receive imports or national inputs with state and federal tax suspension. The final product must be exported without tax payment or internalised with the required tax payment.

CLIA creation was based on the experiments of other countries, mainly the Spanish experience of Logistics and Customs Zones – ZAL, with the difference that in Spain, European Union (EU) directives forbid a fiscally differentiated treatment of goods as the common market permits the free flow of goods and persons among European countries. Table 12.1 compares the activities between CLIAs and dry ports.

---

3   Provisional measures are the most controversial kind of normative act in Brazilian legislation. They have force of law and become effective right after publication in the official gazette; they should be used only in situations of importance and urgency. Available at: http://www.v-brazil.com/government/laws/laws.html#provisional [accessed 16 September 2010].

Table 12.1   Brazil, CLIAs and dry ports: main characteristics of operations

Customs Regimes	CLIAs	Dry Ports
Service Providing Juridical Nature	Private (authorised by SRF)	Public (permission and/ or concession – bidding
Localisation (City/Region)	Defined by private sector	Defined by Government on Bidding
Minimal Structure (area, warehouse, etc.)	Defined by private sector	Defined by Government on Bidding
Special Customs Regime	Permanent	Limited to permission contract, typically, ten years
Services provided related to	Import/Export	Import/Export
Logistics Safety (obligatory requirements)	Site audio and video surveillance and monitoring systems integrated to logistics management system; scanners for containers and pallets; business plan; sampling collected and analyses laboratory.	Electronic Surveillance System

Source: SRF and Tecnologística 2007, adapted by author

As was stated above, the CLIA legal entity was created by a PM which was rejected by the Brazilian Congress due to the matter not being considered "urgent". The Senate subsequently proposed a Law Project: Senate Project Law no. 327 on 13 December 2006, which is until now still under discussion by Senate Commissions.

Nevertheless, in the validity period of PM 320, four firms (see Table 12.2), which had already operated dry ports but whose permission period was almost extinguished, managed to obtain SRF′s authorisation to operate as CLIAs.

This situation deals with different interests around the issue, i.e., firms already operating as dry ports and potential new entrants to the business. Just after PM 320 edition, Sales and Sousa (2006) described the organisations that criticised the PM 320, as saying:

- Brazilian Nation Entities authorised (permitted) dry ports: It considers dry port activities as a public service and, as such, they have to be submitted to biddings and not simply licensed by SRF as the PM 320 established;
- Brazilian SRF Fiscal Auditor Union considered the new rules dangerous to the State by not providing safety to foreign trade operations, as well as by privatising a public service;
- Another dry ports National Association considered the new arrangement (CLIA) as not economically feasible because it requires a public agent control;
- The National Container Terminal Association (ABRATEC) criticised the government's lack of dialogue with the private sector involved.

Table 12.2   Firms considered as CLIAs by SRF during the PM 320/06 validity period

Firms	Certification	Employees	Location	Services Provided
Columbia General Warehousing	ISO 9001/2000	154	Santos (SP)	Banking services, controlled temperature environment, packing and repacking services, industrial processing, customs and other services.
Cragea	None	180	Suzano (SP)	Container operations, palletising, volume controls, container handling, distribution centre, customs and other services
Deicmar	ISO 9001/2000 14000/18000	150	Santos (SP)	All regimes previewed on customs legislation
Mesquita Transports and Services	None	260	Guarujá e Santos (SP)	Industrial processing, assembling, acclimatised warehousing, container operations, customs and other services.

Source: SRF and Tecnologística 2007, adapted by author

The main issue is not these firms´ situations but new projects that could not further develop due to lack of regulation. In September 2010 the Senate Law Project was still being analysed by Senate Commissions.

12.5   Export-Processing Zones (ZPEs)

Today, there are over 3,000 ZPEs or other types of Free Zones in 135 countries around the world, according to Samen (2010). Their objectives are similar to the those established by Brazilian Legislation: (i) enhancement of foreign exchange earnings by promotion of exports of non-traditional manufactured goods; (ii) creation of jobs and generation of income; (iii) improvement of competitiveness of exporters; and (iv) attraction of Foreign Developed Industries with the aim of technology transfer, knowledge spill over, demonstration effects, and backward linkages. (Samen 2010)

ZPEs differ in size and structure, and can be publicly or privately owned or managed. Over the past 10–15 years, as stated by Samen (2010), the number of privately owned or managed zones has grown substantially.

The Brazilian basic legislation dealing with Export-Processing Zones (ZPEs) is Law no. 11,508 from 20 July 2007 that defines ZPEs' fiscal, exchange and managerial regimes; the Decrees no. 6634 from 5 July 2008 that deal with the

ZPEs Special Council (CZPE); no. 6,814 from 6 April 2009 that regulates Law no. 11,508; and SRF Normative Instruction no. 952 from 2 July 2009.

By legislation, ZPEs are defined as free trade (import or export) areas that produce exclusively for exportation. They aim to reduce regional economic imbalances, enhance foreign trade, and promote technological knowledge and Brazil's social and economic development.

Benefits for companies established in ZPEs include import duty and other federal tax exemptions. Companies are also exempt from a special tax charged for financing the marine sector and from financial operation tax. Brazilian legislation distinguishes between ZPEs and free-trade zones (FTZs): enterprises in the latter may sell in the domestic market, while all ZPEs' production must be exported.

ZPEs in Brazil have been regulated by legislation since 1988, but in fact none has been effectively established until the new legislation was promulgated. Today, there are eight authorised by the Federal Government, Assu (RN), Pecem (CE) and Suape (PE) in the Northeast Region; Sen. Guiomard (AC) and Boa Vista (RR) in the North; Aracruz (ES) in the East; Bataguassu (MS) in the Centre-West and Fernandopólis (SP) in the Southeast.

The authorisation does not imply effective operation; it only allows the ZPE implementation actions. In Brazil, ZPEs have to obey a specific and lasting procedure, comprising the following phases:

- ZPE proposals are to be submitted to SRF by State or cities by single or joint proposition;
- The project, including industrial facilities and installations, is analysed by the ZPEs Special Council (CZPE), that is composed of a Ministries collegiate;
- If the proposal is approved, the ZPE is officially created by a Presidential Decree.

The proposal has to demonstrate that the ZPE has adequate logistics, physical area, financial conditions, infrastructure and facilities availability and other required conditions. After the Decree, the ZPE operator has a 12-month period in which to effectively put the ZPE into operation. If this is not accomplished the authorisation decree becomes invalid.

The Brazilian ZPE process is interesting to analyse from two points of view: the first is that of the ZPE critics who point to the lengthy period from the initial legislation to the recent authorisation initiatives. ZPEs are also criticised for opening the possibility of facilitating illegal or undesirable practices, including currency transfer, because of control difficulties by SRF, for example. Another critical argument is that ZPEs could result in unfair competition conditions between firms inside ZPEs and those outside.

The other point of view considers the difficulty or even risks faced by firms intending to embrace this kind of project, as the legislation is not clear on how to deal with public service of this type which are to be managed by private firms.

12.6   Other Special Customs Regimes on Dedicated Areas and Facilitation Procedures

Recently, the Brazilian Government through the Ports Ministry (SEP) has begun to study the conditions required to establish so-called Logistics Activities Zones – ZAL, basically following the Spanish model. These zones are an application of Special Customs Regime areas where logistics value is added to products in order to obtain scale economies, including activities such as kit assembling, cargo consolidation and deconsolidation, packaging, and so on. In South America, the Montevideo ZAL in Uruguay is a noteworthy example. The project has an expected completion period of the first semester of 2011.

Regarding customs facilitation procedures, the Brazilian SRF is proposing on its website a public discussion about the initiative to implement an Authorised Economic Operator – AEO, as has occurred in other countries, which according to the WCO – World Customs Organization (AEO Guidelines) is an intergovernmental organisation focused on customs matters. The proposal is focused on facilitating customs procedures.

In this sense, it is already in practical test and implementation a Government lead by Ports Ministry (SEP), the so-called "Paperless Port". This project is similar to "Single Window" projects that have already been implemented all over the world. SEP finished the project's first phase which involved the documents required to operate a ship at a Brazilian port, involving different government agencies and multiple documents requiring completion. This phase is being tested at Santos Port and it will be extended to other Brazilian ports.

As it can be seen, currently ongoing initiatives have the same aim, i.e., to enhance foreign trade and the competitiveness of Brazilian products in the world market. The different emphases which must be consolidated are logistics services requirements, as seen by SEP, and trade facilitation combined with effective customs control, as pointed out by the WCO:

> Trade facilitation, in the WCO context, means the avoidance of unnecessary trade restrictiveness. This can be achieved by applying modern techniques and technologies, while improving the quality of controls in an internationally harmonised manner (WCO 2010).

So, it can be concluded that both agencies are on a similar path and they share the same belief, as cited by WCO 2010:

> Trade facilitation is one of the key factors for economic development of nations and is closely tied into national agendas on social well-being, poverty reduction and economic development of countries and their citizens (WCO 2010).

The initiatives could be implemented in parallel, i.e., the facilitation of customs procedures and Special Customs Regimes on dedicated areas are not mutually exclusive. In fact, if the formal procedures are facilitated, they will improve the logistics aspect of a project and the discussion will be which areas or regions have the best logistics advantage or infrastructure to receive this kind of business. A basic assumption is that the private sector should be the main agent. This can be considered as another relevant issue to be dealt with and clarified in the regulation.

Assuming logistics and infrastructure as determinant issues, RMBS naturally appears as a potential candidate to receive Special Customs Regime sites, as has already occurred with bonded warehouses in the Santos Port Primary Zone.

12.7   RMBS Perspective and Potential

RMBS was created in 1996, by the State Law no. 815 of 30 July, comprising 2,373 km^2 divided among nine cities: Bertioga, Cubatão, Guarujá Itanhaém, Monguagá, Peruibe, Praia Grande, Santos and São Vicente with 1.4 million inhabitants and a seasonal population of 4.9 million (Assumpção et al. 2009). Figure 12.2 shows its location in São Paulo State and the RMBS's cities.

The metropolitan area has as its main characteristics: political, social and economic integration, and an intense population migration among its cities. Common or integrated solutions to problems are not considered from a regional aspect. Accessibility, water and sewage sanitation projects are poignant examples.

In the specific case of RMBS, activities related to ports are doubtless one of its main tasks. The Santos Port is located in three RMBS cities, Santos, Guarujá and Cubatão, and its hinterland extends to the greater part of Brazil and even to

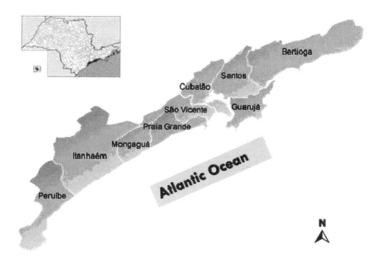

Figure 12.2   RMBS Cities
Source: Adapted by Author.

other South American countries such as Bolivia, Paraguay, Argentina and Uruguay (Mercosur countries).[4]

The Santos Port hinterland includes the greater part of Brazilian agribusiness exports and it is close to the São Paulo industrial area, the most important in Brazil. Santos Port is well served by land transport infrastructure both by railway (the private operators MRS Logistics S/A. and ALL – Latin America Logistics S/A.) and highway (Anchieta-Imigrantes System, also privately operated). The main Southeast Region international airport (Cumbica in Guarulhos city in the East of Great São Paulo Metropolitan Region) is located no more than 100km from the port.

These characteristics point to RMBS as a natural candidate in which to locate a Special Customs Regime Area (Dry Port). It is important as well to consider environmental aspects, which are very critical in an RMBS, especially for areas located by the sea and in light of Santos Port's estuary characteristics. New expansion projects are dealing with this issue, in spite of costs and time requirements, considering the environmental issue as not simply a restriction or legal enforcement, but as a strategic dimension of their business.

Another relevant issue for RMBS is that its population has a strong seasonal employment resulting from the summer holidays, which attracts a huge quantity of tourists to its beaches. Dry Port projects could mitigate this problem by providing steady jobs all over the year. The RMBS cities' projects in that direction are described in the following section.

12.7.1   Cubatão City Council

The City Council has implemented actions to enhance port related activities in the city, such as fiscal incentives, city tax exemptions, and access facilities to local areas, in order to attract private projects. Cubatão is the main access to Santos Port left bank, where the main container terminal is located. In the city, truck parking lots have also been established in order to regulate truck flows in and out of the Santos Port Primary Zone.

12.7.2   Guarujá Seaside Resort[5]

Guarujá City Council considered a project to implement a CLIA in a 4.1 million m² area located at Cônego Domenico Rangoni Roadway, the main link between Cubatão and Santos Port left bank. The project foresaw a container warehousing

---

4   Mercosur: International organisation consisting of Argentina, Brazil, Paraguay, and Uruguay, as well as other associate members in Latin America. The organisation mandates lowering of tariffs and other trade barriers, with an eye towards eventually eliminating restrictions on the movement of capital, labour, goods and services. It aims to increase trade by and among South American countries.

5   Seaside resort: State cities classification that allows them to receive special treatment regarding tourism resources. They are defined and regulated by specific São Paulo state law.

facility, truck parking lot, industrial installation, and terminal for liquid bulk and refrigerated cargoes. This project did not go further due to CLIA PM's extinction. The city has an airport expansion project, on a former Brazilian Air Force airfield, that it is already being used by PETROBRAS as logistics support for its oil exploitation platforms in Santos Basin.

12.7.3   Itanhaém Seaside Resort

The city has a state-owned airport with a 1350m-long and 30m-wide runway. Today it is used for passenger transportation and it can receive medium-size airplanes such as the Boeing 737-400 and Fokker 100. This airport is also used as a helicopter landing field by PETROBRAS for logistics support to its offshore oil and gas operations. The City Council has developed a project to implement a Special Custom Regime Area, but to date without any success.

12.7.4   Praia Grande Seaside Resort

The City Council has developed a project to implement an International Cargo Airport that has already had technical feasibility studies approved, as has been announced by the Business Relationship Secretary and private groups interested in the enterprise. The project is planning a Special Customs Regime Area (Dry Port) and areas to industrial installations and facilities. In personal interviews with an executive of the interested firm, it has been pointed out that the project, despite being considered technically and economically feasible, may not proceed due to a lack of firm regulation and the severe requirements of the present legislation.

It has been said: "This matter can hardly be solved until the next presidential election [scheduled for] next October. Even so, we have to wait [for] the new government to be installed. Probably, we´ll have news in the mid of next year".

12.7.5   Santos Seaside Resort

Recently, Santos City Council appointed a Port and Maritime Issues Secretary and the Secretary is Chairman of the Port Authority Council (CAP), indicating its port city character. In this sense, this Council is participating in Santos Port expansion discussions and its main concern is the access infrastructure to the Port right bank. Santos' restriction is the lack of open areas required to install Special Customs Regimes projects. Its advantage is its urban structure and skilled workforce, which make the city a specialised services provision centre. An example is the PETROBRAS decision to install its administrative headquarters for a pre-salt layer oil and gas exploration project in Santos.

12.7.6 São Vicente Seaside Resort

São Vicente City Council has recently implemented an Industrial Project located in a dedicated area, in order to attract private sector investment. The project has also planned an area for a Special Customs Regime enterprise.

As is becoming apparent, RMBS city authorities are engaged and have reached agreement about the activities of relevance to the region; however, for their actions to be effective, they are depending upon federal legislation. The presidential election period will delay a definite solution.

12.8 Conclusion and Remarks

Foreign trade and international logistics activities can be considered as part of RMBS's economic inclination, due to Santos Port's location, historical development, and infrastructure services as well as its skilled workforce. The efforts of RMBS´s city councils are examples in this sense.

Regarding Special Customs Regimes, Brazilian regulation is very complex and it can be considered to be in a transition period towards facilitating and enhancing foreign trade procedures. This can be summarised in two main dimensions: the logistics aspects, focusing on location sites, regarding basic infrastructure close to ports or airports, accessibility and legal regulation; and the other dimension, also based on legal regulation, but not necessarily related to a specific site or location. Both are related to authorisation processes and are based on IT support.

In Brazil, as far as this study is concerned, the Special Customs Regime Dedicated Areas are dependent upon effective regulation solutions, with ZPE projects being in a more concrete stage at present. Facilitation procedure initiatives are ongoing in different phases and governmental agencies have the agreed objective of avoiding trade restrictiveness, without losing effective control on merchandise flows.

The final conclusion is that despite the name that a dedicated area with special customs regime will receive (Dry Port can be a generic one), the perceptions of its importance and its potential role are shared by governmental agencies with a vision of enhancing foreign trade as part of a more general objective towards poverty reduction, and job and wealth creation opportunities that are open to all the people.

12.9 References

ABRAZPE 2007. Associação Brasileira de Zonas de Processamento de Exportação. *Presentation 2007* [online]. Available at: http://www.abrazpe.org.br [accessed 5 June 2008].

Assumpção, M.R.P., Alves, A.G. and Robles, L.T. 2009. *Implantação de Áreas Dedicadas a Regimes Aduaneiros Especiais na Região Metropolitana da Baixada Santista: Zonas de Processamento de Exportação e Centro de Logística e Indústrias Alfandegadas: Proceedings of XII Semead – Entrepreneurship and Innovation.* São Paulo: XII SEMEAD.

Brazilian Federal Senate 2010. *Acompanhamento de Matérias Legislativas.* Available at: http://www.senado.gov.br/atividade/materia/Detalhes.asp [accessed 5 September 2010].

Brazilian Federal Revenue Secretariat – SRF 2010. *Operadores Econômicos Autorizados* [online]. Available at: http://www.receita.fazenda.gov.br/aduana/ OperEconAutorizados/default.htm [accessed 10 September 2010].

Cardoso, F. 2007. Muitas dúvidas e poucas certezas. Portos secos: Mudanças na legislação confundem o mercado. *Revista Tecnologistica*, 12(138), 74–85.

Ciesa, S.A. 2007. *Divulgação das Indústrias Alfandegadas* [online]. Available at: http://www.industriasalfandegadas.com.br/noticia.asp?sec=5 [accessed 3 October 2007].

CODESP 2010. *Relatório da Companhia Docas do Estado de São Paulo* [online]. Available at: http://www.portodesantos.com/ [accessed 20 August 2010].

Faro, R. and Faro, F. 2007. *Curso de Comércio Exterior.* São Paulo: Atlas.

Lourenço, M. 2008. *Portos Secos: Vantagens* [online]. Available at: http://www. netcomex.com.br [accessed 10 September 2008].

Ricupero, R. and Oliveira, L.V. 2007. *A Abertura dos Portos*, 1st Edition. São Paulo: Senac.

Sales, A. and Sousa, V. 2006. Comércio Exterior: Novas regras geram polêmica nos portos secos. *Revista Tecnologística*, 11(132), 44–53.

Samen, S. 2010. *Export Development, Diversification and Competitiveness: How Some Developing Countries Got it Right.* World Bank Institute – Growth and Crisis Unit. Draft for Comments, March, 10 [online]. Available at: http://blogs. worldbank.org/growth/Export-Diversification-Competitiveness-Paper-Samen [accessed 9 September 2010].

Uhy Report 2010. *Doing Business in Brazil* [online]. Available at: http://www. uhy.com/media/PDFs/doing_business_guides/Doing%20Business%20in%20 Brazil.pdf [accessed 14 September 2010].

WCO 2010. *Report 2010* [online]. Available at: http://www.wcoomd.org/sw_ overview.htm [accessed 21 September 2010].

WTO 2004. Brazil Continues to Liberalize: Further Steps would Benefit the Economy and World Trade. *Trade Policy Review*. Report by the Secretariat [online]. Available at: http://www.wto.org/english/tratop_e/tpr_e/tp239_e.htm [accessed 15 September 2010].

Chapter 13

Potential for Logistics Zones Development: Chile as a Case Study

Erick Leal Matamala,[1] Gabriel Pérez Salas and Ricardo J. Sánchez

13.1   Introduction

Although the most developed logistics zones are located in the main global economies, the maturity of these markets and the increasing integration of secondary economies into the global logistics and supply chain add a new geographical perspective to the developments of logistics zones. Both established and new multinational companies (MNCs) are diversifying their business and are looking for new partnerships in new geographical markets. At the same time, global logistics operators (GLOs) are expanding their door-to-door services to meet with the new geographical scope of global supply chains. In such a scenario it is valid to hypothesise that typical logistics poles developed in the main global economies will require a counterpart in emerging economies in order to support the geographical integration of logistics and manufactures networks. Thus, MNCs and GLOs are looking for new places and new partners to allocate their production and distribution activities, generating opportunities for local economies to develop logistics poles according to their respective realities.[2]

In line with the previous arguments, certain developments can be considered as evidence in the South American case. For example, two global port terminal operators (and inland terminal operators), Hutchinson Port and DP World, have manifested interest in expanding their operational branches in Ecuador and Peru respectively. From an infrastructure perspective, local governments have put emphasis on improving the countries' port systems and highway networks. Furthermore, the national interests in developing the logistics sector have been supported by transnational initiatives, particularly the South American Regional Infrastructure Integration Initiative (IIRSA).

---

1   Corresponding author.

2   At a theoretical level, such a hypothesis is founded on the works of Slack (1990) and Notteboom and Rodrigue (2008). In the former, Slack identified the phenomenon of multimodality as the main inductor of the development of inland distribution centres in the North American market. In the new global context, Notteboom and Rodrigue argue that logistical integration is the main factor that puts increased pressures on inland development.

This chapter focuses on the case of Chile. Chilean economy generates the largest volume of containerised traffic and trades the highest value in international trade on the West Coast of South America (WCSA): almost three million TEUs and 42 per cent of the total value of international trade on WCSA respectively (World Trade Organization WTO 2010). At the same time, the Logistics Performance Index 2010 (LPI, World Bank 2010) ranks the country highest of those on the West Coast of South America, with a score of 3.09.

Thus, the objective of this chapter is to determine the potential of logistics zone development in Chile. The authors assume that such a potential can be identified by observing the site selection of logistics operators. The methodology includes an econometric model revealing the main variables shaping the decision process of site selection, and a further cluster analysis is applied to rank the 49 Chilean provinces in terms of "logistics relevance". This methodology is an alternative to the approaches based on perceptions and operational research models. The results deliver empirical evidence to existing literature that traditionally focuses on the main global economies and gives less attention to secondary markets.

The authors state that some of the proxy variables should be replaced in the future, in order to validate the variables and the measuring methods used in the chapter due to a lack of primary information (especially referring to the number of logistical companies, to the public services approached by the Agriculture and Livestock Service (SAG) Offices, and the working capacity, estimated through the number of graduates).

13.1.1   Literature Review

"Logistical platforms", "distribution centres", "logistical zones" or "logistical parks" are usual denominations for transport infrastructure, which concentrates national and international cargo flows through one or more transport modes, giving the opportunity to provide additional services related to transport, distribution and logistics activities, involving both private and public participation and ensuring the participation of any company with conditions for competing under the free market rules (Europlatform 2004). Table 13.1 lists principal authors and articles that have investigated the development of logistics infrastructure, i.e. the development potential of logistics zones and the factors determining such potential.

A first and basic approximation to define logistics zones relates to the development of international hub port infrastructure. In this case, the port itself can be understood as a logistics zone or "cluster". Notteboom (1997) discusses the role of technology, organisational issues, accessibility and public policy in relation to a port's capacity to concentrate cargo. Slack (1990), van Klink and van den Berg (1997) and Mc Calla (1999) demonstrate that the promotion of inter modal infrastructure is an effective strategy to meet with the status of logistics platforms. Similarly, Hoffmann (2000) argues that international trade volume, location, local market size, technological development, scale economies and freight rates are important variables that do justify such infrastructure. The work of

De Langen (2002) explains how agglomeration economies are the main inductor for the development of the maritime cluster in the Netherlands. A more specific explanation is given by Ducruet (2005) and Ducruet and Lee (2006). They utilise the concepts of centrality and intermediacy at a global scale to identify port cities with potential for hub activities. Finally, Tongzon (2007), for a set of international manufacturing firms, found that incentives for foreign investors, infrastructure development, domestic economy and political environment, government policy and political regulation, are the most important factors shaping their decision regarding location.

Along the same lines of hub development, the hub location problem in the context of operational research is another branch offering a set of variables which define the location decisions of firms. Although the literature addresses the problem from different approaches, it is possible to identify common variables determining the optimal solution. Rodriguez et al. (2007), Kara and Tansel (2001) and Campbell (1996), just to mention a few from the long list of articles, include variables such as: volume, unit cost, distance and number and capacity of hubs. In the specific case of the port industry, Aversa et al. (2005) found as an optimal solution for South America the port of Santos in Brazil and as a second best solution the port of Buenos Aires in Argentina. Following the work of Campbell (1996), they determined this solution through: costs (port, vessel daily, container, road transport) nautical and road distance, vessel speed, container flow direction and economies of scale, among others.

A third point of view of the problem is addressed through the Foreign Direct Investment (FDI) location determinants. Oum and Park (2004) studied the location determinants of regional distribution centre in the northeast Asian region. They found that the most important variables influencing the manager's decision to invest were geo-location, transport linkage and accessibility. At the same time, market size and growth potential were the second most important ones. Port, airport and intermodal transport facilities were set at a third level of importance; while skilled labour force, labour quality and labour peace occupied the last place of the ranking. Along the same lines, Lu and Yang (2007) evaluated the importance of market size, cost, infrastructure and political factors with logistical investment intentions. The findings revealed the importance of the first three factors as the main drivers of investment intention, with a minor importance in political dimensions. More specifically, they found that market factors were explained by economic growth and market size; cost factors by cost of land, labour cost and corporate tax index; and infrastructure factors by communication systems, skilled labour force and transport linkage. Furthermore, the lesser importance of the political factor was explained by political stability, security and efficiency of government administration.

Finally, Riemiené and Grundey (2007) establish an operational hierarchy that includes three levels of logistical places according to complexity. The first level considers the concept of warehouse and distribution centres. Usually at this stage there is only one transport mode (road), and the main function is storage (in the case of warehouse) and flow management for the final customer (in the case of distribution centres). The second level in the hierarchy includes transport terminals, logistic centres and logistic villages. The main difference with the first order facilities lies in the transport modes involved and their respective functions. At this level, two or more transport modes concentrate either transshipment traffic or cargo, which have big and dense markets as their final destination, giving the opportunity to offer other related services. The third level consists of logistical nodes, which are places incorporating a complex network of logistic platforms, usually in a broad scope facing different markets and their respective needs. An interesting widening of this hierarchy is given by Leal and Pérez (2009). They add risk, organisational issues and the public sector as important variables to take into account in the implementation process of this kind of infrastructure.

Additionally, the hierarchy established by Riemiené and Grundey (2007) allows the incorporation of small and remote economies where the role of logistics platforms may be crucial at regional or local scale. From a methodological point of view, conceptual development is useful as an initial approach; at the same time, operational research tools do not consider important aspects such as political issues or some specific characteristics of human behaviour, even though they are very useful when rationalising a decisional model. Declared preferred techniques modelled with psychometric tools (such as Lu and Yang 2007) add evidence based on what managers "say". However, people do not always do what they say, thus, an important challenge lies in revealing preferred techniques. Furthermore, existing literature focuses on developed economies, giving less attention to developing countries, so new methods in incipient markets turn out to be an interesting challenge in this research line.

The rest of the chapter is organised as follows: Section 13.2 describes the methodology, Section 13.3 depicts results and discussion; and the conclusion and suggestions for further research are included in Section 13.4.

* * *

Table 13.1   **Matching authors and variables shaping logistical potential**

	Slack (1990)	Notteboom (1996)	Zinn (1996)	Campbell (1996)	van Klink and van den Berg (1997)	Mc Calla (1999)	Hoffmann (2000)	Kara and Tansel (2001)	de Langen (2002)	Oum and Park (2004)	Ducruet (2005)	Aversa et al (2005)	Ducruet and Lee (2006)	Tongzon (2007)	Rodríguez et al. (2007)	Lu and Yang (2007)	Notteboom and Rodrigue (2008)	Leal et al. (2009)
Investment intention (endogenous)																✓		
Location (endogenous)				✓	✓		✓	✓		✓		✓	✓	✓			✓	✓
Transport (land accessibility)	✓	✓	✓	✓	✓	✓	✓	✓		✓	✓	✓	✓	✓		✓	✓	✓
Maritime accessibility	✓	✓	✓	✓	✓	✓	✓	✓		✓		✓	✓		✓	✓	✓	✓
Market size (GDP)		✓	✓			✓	✓		✓	✓		✓	✓		✓	✓	✓	✓
Market growth		✓					✓		✓	✓			✓		✓	✓		
Port facilities	✓	✓	✓		✓	✓				✓			✓			✓	✓	✓
Airport facilities				✓		✓						✓						
Intermodality	✓	✓	✓			✓	✓	✓	✓	✓				✓		✓	✓	✓
Labour market										✓				✓			✓	
Cost	✓		✓	✓	✓	✓		✓				✓		✓	✓	✓	✓	
Political factors															✓	✓		✓
Public services		✓								✓	✓			✓		✓		✓
Taxes																✓		
Traffic volume	✓	✓	✓	✓	✓	✓	✓	✓	✓	✓	✓	✓	✓	✓	✓		✓	✓
Distance	✓	✓			✓				✓		✓	✓	✓		✓		✓	✓
Trade flow direction				✓	✓							✓	✓	✓	✓		✓	✓
Private operators							✓		✓								✓	✓

Source: Authors

13.2   Methodology

The research design is depicted in Figure 13.1. Literature review and expert interviews have been the main sources for defining the kind of information required. In a second stage, a set of variables representing the problem being studied have been selected according to the availability of public information and previous literature review. The study unit has been decided according to two criteria: information availability and sample size. Concerning specific tools, an econometric model has been implemented to reveal the structure of the problem and the relative importance of each variable. Based on this, cluster techniques have been implemented to reveal the potential for logistic development of each province.

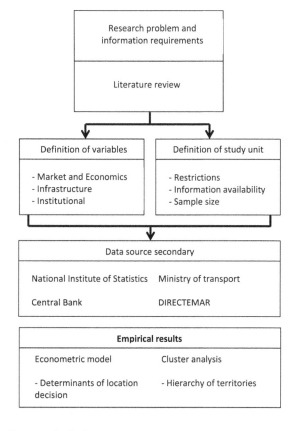

Figure 13.1   Research design

13.2.1 Variables and Study Unit

Table 13.2 depicts a summary of the main variables defining the problem according to the literature review. The set of variables can be given a preliminary classification as market/economic variables, labour, infrastructure, location and public services influencing the potential for logistical development.

Potential for logistical development The number of companies associated with the Logistics Association of Chile (ALOG) with operations in each province is used as a proxy for measuring the potential for logistical development. The business unit level has been defined as the study unit in order to avoid the problem of mergers and acquisitions. ALOG includes the most important global, regional and local logistics operators, including maritime and land transport, freight forwarders, logistics service providers, infrastructure service providers and port terminal operators. The companies that are part of ALOG represent around 90 per cent of the market share in the Chilean market (www.alog.cl).

Market/economic variables Gross domestic product, export volume and other relative measures are not available in the National Account System at a provincial level. Therefore, population density and income data from the CASEN survey will be used as proxies for this dimension.

Labour variables A strong labour market with availability of skilled labour force is a key element to reach high degree of productivity. Thus, the numbers of technicians and professionals who graduated during 2007 are used as two indicators for labour market development. The data was obtained from the statistics system of Education Ministry.

Infrastructure variables In absolute terms, longitude of roads and railway systems give a first approximation in relation to the availability of transport infrastructure inside the territory. Additionally, the number of port terminals is used as a proxy for maritime accessibility. The data set is obtained from Ministry of Public Works and General Board of Maritime Territory and Merchant Marine (DIRECTEMAR) respectively.

Location variables The shorter distance to a port or main urban centre implies good candidacy for regional or local status. Both distances come from official maps and complemented with Google Earth tools.

Public services Although the literature review includes political stability as an important variable, the availability of public service related to logistics business is an additional set of important variables shaping the logistical profile of territory. The numbers of Customs offices and SAG offices were used as proxies for availability of public services.

Table 13.2   Selected variables

Variable	Variable Name	Source
Potential for logistics	Number of ALOG operators	Web pages of enterprises combined with the information from yellow pages directory
Market	Monetary income Chilean pesos at 200x	CASEN survey 2006
	Market density: population divided by squared kilometres of surface of the province	Census 2002
Labour variables	Availability of professional labour force: number of professional graduated during 2007	Ministry of education
	Availability of technical labour force: number of technicians graduated during 2007	
Infrastructural variables	Longitude of motorway system in kilometres	Ministry of public works, Ministry of Transport
	Longitude of railway system in kilometres	State enterprise of railway (EFE)
Location variables	Distance to the main export port in kilometres	Google earth
	Distance to the main urban centre in kilometres	
Public services variables	Number of custom offices Number of SAG offices	Custom web page SAG web page

Source: Authors

All the previous variables have been calculated at a provincial level;[3] the reason behind this decision lies on data availability and methodological issues. With regard to the former, data available at Central Bank, as well as INE and DIRECTEMAR, is only organised at a regional level. And concerning methodological issues, the reduced number of regions (13) puts several limitations on the statistical techniques, which require variability and a minimum number of observations to meet with the assumption of normality and consistency of estimations. Thus, the province level appears to be the most convenient way to collect the data. Annex 1 depicts the 49 provinces in the study.

---

3   Chilean territory is politically organized in regions, which account for a number of provinces whose basic organizational unit is the municipality.

13.2.2 Modelling the Problem

A first step for modelling the problem concerns an econometric approach in order to reveal both the statistical significance and the effect of the factors on the potential of logistics zones for the specific Chilean case. To do that, the variables described above are introduced in the following specification:

$$Lg = \alpha + \beta_{m_1} m_1 + \beta_{m_2} m_2 + \beta_{l_1} l_1 + \beta_{l_2} l_2 + \beta_{f_1} f_1 + \beta_{f_2} f_2 + \beta_{f_3} f_3 + \beta_{c_1} c_1 + \beta_{c_2} c_2 + \beta_{p_1} p_1 + \beta_{p_2} p_2 + e \quad (1)$$

Where:
Lg = Number of logistics operator business units in the ALOG association
m_1 = Population density; m_2 = Income
l_1 = Number of professionals graduated from Universities and institutes; l_2 = Number of technicians graduated from Universities and institutes
f_1 = Linear kilometres of motorway; f_2 = Linear kilometres of railway; f_3 = Number of port terminals
c_1 = Linear kilometres from the nearest port; c_2 = Linear kilometres from Panama Canal.
p_1 = Number of SAG offices and p_2 = Number of Customs offices
α = constant term; $\beta_{m_i}, \beta_{l_i}, \beta_{f_i}, \beta_{c_i}, \beta_{p_i}$ are parameters of the vectors of market, labour, infrastructure, cost (location) and public services.
Finally, e represents the variables not included in the specification.

The second stage concerns the application of clustering techniques, namely, the K-means algorithm upon the resultant variables from the first stage.

13.3   Results

13.3.1 Econometric Approach

Model A (Table 13.3) shows a first regression under the ordinary least squares technique (OLS). Albeit the very high level of fit (R-squared = 0.971), the increased variance factor (IVF) reveals several problems of collinearity (IVF > 10). To avoid misinterpretations given by biased coefficients, the model has to be refined. Specialised literature gives some solutions such as increasing sample size or using indicators related to one or more variables involved in the problem. According to the IVF depicted in Table 13.3 (Model A), market and labour variables present several problems of collinearity. Note that it does not reveal problems of erroneous specification necessarily; on the contrary, from a theoretical point of view, there is a very acceptable strong relationship between them. First of all, in a highly developed market, the probability of finding a skilled labour force should also be high. And consequently, a skilled labour force would influence the market development.

Given the previous argument, the proposed solution is to implement an indicator or a latent variable representing the "market development". To implement this latent variable, factor analysis through principal components analysis is run on the four observed variables: m_1, m_2, l_1 and l_2. The results on Table 13.4 depict only one Eigen Value greater than 1 (3.914), at 97.84 per cent of extracted variance. The scores of each province for this factor represent the new variable denoted by m_{pca}, and the new specification for models B and C in Table 13.3 is given by (2):

$$Lg = \alpha + \beta_{m_{pca}} m_{pca} + \beta_{f_1} f_1 + \beta_{f_2} f_2 + \beta_{f_3} f_3 + \beta_{c_1} c_1 + \beta_{c_2} c_2 + \beta_{p_1} p_1 + \beta_{p_2} p_2 + e \tag{2}$$

Where $\beta_{m_{pca}}$ is the newly introduced parameter.

In Table 13.3, Model B reveals how well the implemented solution works. Every IVF in Model B shows evidence of very low levels of collinearity. Furthermore, it is possible to make a first round of basic interpretations. First, it is clear that the model does not lose fit, R-squared and the statistical significance of the F-value reveals that total variance is explained by the coefficients. Second, partial analysis of every coefficient reveals how collinearity affected the first model. The constant term is now positive and significant; the impact of f_3 given by β_{f_3} is lower, while the impact of f_2 given by β_{f_2} is higher. Although the sign of β_{f_1} does not change, it is notably higher.

At the same time, if the non-significant regressors are eliminated, Model C reveals the final structure of the problem. Specifically, the highest value of $\beta_{m_{pca}}$ and the smallest value of p reveal that "market development" (m_{pca}) is the most important determinant of the number of ALOG operators in the Chilean market. The same parameters indicate that the second most important variable is f_3 ("number of port terminals"). Concerning the less important variables, the significance of β_{f_2} makes it possible to keep this variable despite its low relative value. In addition, although the p-value of β_{f_1} is on the limit, the variable has been kept.

These results are in line with previous literature. Hoffmann (2000), De Langen (2002) and Lu and Yang (2007) identified "market characteristics" as the main determinant of logistic zone development. Actually, the population density and the greater income in a province generate enough volume of cargo to gain the advantage of scale and agglomeration economies. Concerning the labour market, the availability of skilled labour force is crucial to ensure effective knowledge transference in a context where organizational capabilities take increasing importance (Leal et al. 2009). According to the specification given in (2), increments of one point in m_{pca} give incentives for the allocation of fifteen additional companies.

In the case of "number of port terminals", the literature on hub port development argues that port terminals are indeed inter modal transport nodes, linking maritime and land transport modes. This implies that port infrastructure attracts a great volume of cargo flow given the scale economies of maritime mode. Additionally, port terminal operators depend on other logistic services such as stowing, cargo

Table 13.3 Results of the econometric approach

	Model A				Model B				Model C			
	Non-Standard coefficient B	Standard coefficient Beta	Sig. (p-value)	IVF	Non-standard coefficient B	Standard coefficient Beta	Sig. (p-value)	IVF	Non-standard coefficient B	Standard coefficient Beta	Sig. (p-value)	IVF
α	-1.240		0.399		2.807		0.066		4.170		0.000	
m_1	0.011	0.240	0.127	30.49								
m_2	0.000	0.244	0.249	56.06								
l_1	0.013	1.361	0.000	118.10								
l_2	-0.021	-0.886	0.001	74.09								
f_1	-0.004	-0.013	0.734	1.79	-0.023	-0.080	0.061	1.48	-0.023	-0.077	0.050	1.31
f_2	0.009	0.069	0.053	1.55	0.013	0.094	0.018	1.25	0.013	0.095	0.010	1.11
f_3	1.345	0.166	0.000	2.06	0.784	0.097	0.020	1.39	0.915	0.113	0.002	1.10
c_1	0.001	0.025	0.495	1.70	0.000	0.003	0.948	1.64				
c_2	0.000	-0.002	0.956	1.43	0.002	0.014	0.731	1.38				
p_1	0.083	0.010	0.809	2.06	0.283	0.033	0.476	1.86				
p_2	0.352	0.012	0.768	2.01	1.218	0.041	0.383	1.83				
m_{pca}					14.817	0.982	0.000	1.47	15.052	0.998	0.000	1.33
R-squared	0.971				0.954				0.951			
Corrected R-squared	0.963				0.945				0.946			
Sum of squared residual	10,608.320				10,417.616				10,383.900			
Sum of squared regressors	313.802				504.506108				538.222425			
Sum of total squared	10,922.122				10,922.122				10,922.122			
F	113.701				103.245691				212.222			
Significance	0.000				0.000				0.000			

Source: Authors

services, container and other shipping services, which attract the allocation of other supply chain firms. According to the specification given in (2), for every additional terminal, the allocation of one additional company is hoped for.

Regarding the less important variables, a first intuitive explanation is drawn from the works of Slack (1990), Van Klink and van den Berg (1997) and McCalla (1990). They argue that multimodality given by railway and motorway development gives an incentive to the hub status and then, the concept of "port cluster". However, the negative sign of β_{f_1} challenges such a hypothesis indicating a different dynamic into the Chilean case. Actually, some important ports account an increasing importance of certain market segments characterised by big volumes and a high degree of geographical concentration of production. In this scenario, motorway loses competitiveness given the importance of scale economies, and shippers take the advantage of railway. A good example is the forestry sector at the south of Chile, where 99.9 per cent of paper pulp is transported by railway.

13.3.2   Cluster Analysis Approach

The K-means algorithm is used under three different configurations, and the significance of the F-measure is applied to check how well separated the resultant clusters are. Table 13.5 depicts the significance of the F-measure for every variable under three different values for K. With $K=3$ and $K=4$, significance of the F-measure improves about $K=2$. For $K=4$, the separation among clusters is better, mainly for "motorway" and "port terminals", while the improvement for "railway" is only up to $K=3$.

Table 13.4   Factor analysis for the "market development" variable

Variables	Initial	Extracted
m_1	1.000	0.964
m_2	1.000	0.985
l_1	1.000	0.989
l_2	1.000	0.976
Eigen Value		3.914
% variance		97.839

Source: Authors

Table 13.5   Statistical significance of distances

Variable	K means		
	Significance of F at $K=2$	Significance of F at $K=3$	Significance of F at $K=4$
M_1	0.000	0.000	0.000
M_2	0.000	0.000	0.000
L_1	0.000	0.000	0.000
L_2	0.000	0.000	0.000
F_1	0.005	0.004	0.001
F_2	0.337	0.176	0.238
F_3	0.571	0.038	0.013
Lg	0.000	0.000	0.000

Source: Authors

Panel A in Table 13.6 depicts the characteristics of every cluster for $K=3$ and $K=4$, while Panel B shows cluster membership for provinces. The result in cluster one is consistent with an econometric approach, the "market development" variable being the only one capable of explaining such potential, as in fact there are no terminals or railways inside it. The results in cluster two reveal the importance of "port terminal infrastructure", as for $K=3$ it appears with an average of three port terminals against zero in cluster one. At the same time, if $K=4$ the configuration is clearer than before. Cluster two shows an average of four terminals against only one terminal for cluster three and four. In the case of cluster three, railway infrastructure plays an important role against a weak participation of market and port variables; in fact, when $K=4$, cluster three is calculated with an additional infrastructure of 56 per cent over cluster two, while the difference in motorway is almost zero. This last result is consistent with the low significance of the motorway parameter in the econometric approach, whose p-value was on the limit (0.05). Concerning the specific cases eliminated from cluster two when $K=4$, such status changes mainly because of weak development in population and income, although the specific case of Antofagasta deserves special attention. Indeed, Antofagasta included 17 ALOG operators, similar to Concepción, and additionally accounts for four port terminals, similar to Valparaíso. The reason behind its elimination from cluster two lies in the importance of market development, where this province presents a weak ranking on both population density and income.

Finally, if $K=4$ is taken as the definitive configuration, Panel B in Table 13.6 shows that Santiago is the province with highest potential for logistic zone development with 101 ALOG operators. It is followed by Valparaíso, Cordillera and Concepción in cluster two, with an average of 14 ALOG operators. Clusters three and four show a weak potential, with five and two ALOG operators on average, respectively. These results show again the great importance of the "market development" variable. Indeed, Santiago has no port infrastructure and

Cordillera is classified jointly with Valparaíso and Concepción, coastal provinces with a high development in port infrastructure.

The previous results may be interpreted in the light of both the theoretical aspect given by the previous literature review and the current situation of Chilean market. In the former, "market development" is clearly the most important variable and it explains why Santiago and Cordillera are classified in cluster one and two respectively, even when they do not have multimodal infrastructure. At the same time, logistics zone development has been mainly associated with coastal location where port infrastructure takes a relevant role. In this way, Valparaíso and Concepción in cluster two appear as consistent results. Finally, inland location for logistics zone development is sustained mainly on the existence of railway or inland navigation channels, and this argument is consistent with the results in cluster three, where the most of the provinces are inland ones with relevant levels of railway infrastructure. Here, it is important to mention that the Chilean government withdrew its subsidies in 1979 and the private participation started only in 1995, which mean 15 years of losing competitiveness compared with other modal choices such as motorway.

Along the same line, the statuses of Valparaíso and Concepción have been favoured with an aggressive policy of global integration where port privatisation has been a fundamental base (Law no. 19.542, implemented in 1998). Although it provides some limitations to avoid monopolistic behaviour, this Law allows the mono-operation of public terminals giving to private firms the conditions to reach scale economies and high levels of productivity (See Doerr and Sánchez 2006). At the same time, maritime concessions have not been restricted to private port operators, allowing the birth of important private port companies, related mainly to the forestry and mining sectors, for example, Port of Lirquén, Port of Coronel, and a later concession at the Concepción bay: The port of Los Reyes.

Table 13.6   Clustering results

Panel A: Means for K=k	Means for each cluster at K=3			Means for each cluster at K=4			
	1	2	3	1	2	3	4
Ingreso CASEN $MM	4.760.000	462.000	92.498	4.760.000	571.000	236.462	58.434
Pop density	2.299	135	37	2.299	231	80	20
Labour Technic	4.353	471	45	4.353	733	117	31
Labour Prof	10.996	981	89	10.996	1.525	294	38
Terminals	0	3	1	0	4	1	1
Km motorway	187	75	39	187	78	72	29
Km railway	0	176	99	0	100	156	90
ALOG	101	11	2	101	14	5	2
Panel B: Cluster membership of provinces							
	Santiago	Antofagasta	Iquique	Santiago	Valparaíso	Arica	Iquique

Valparaíso	Arica		Cordillera	Antofa-gasta	Parina-cota
Cordillera	Putre		Concep-ción	El Elqui	El Loa
Cachapoal	Calama			Maipo	Tocopilla
Concep-ción	Tocopilla			Talagante	Chañaral
Cautín	Chañaral			Cachapoal	Copiapó
	Copiapó			Curico	Huasco
	Vallenar			Talca	Limarí
	-			Ñuble	Choapa
	-			Biobío	-
	-			Cautín	-
	-			Valdivia	-
	Tierra del fuego			Llanqui-hue	Tierra del fuego

Source: Authors

An important finding for the specific Chilean case is related to the specific role that logistics platforms take in the whole logistics chain. "Coastal zones" support logistics activities related to the port and maritime segment such as warehousing, container maintenance and public services related to customs and public health. At the same time, "inland zones" with high levels of market development indicate a potential for services related to the shipper, this means, cargo and managerial services, such as cargo consolidation, warehousing, packaging, transport and information technology services.

13.4   Conclusions

The establishment of a global supply chain imposes to emerging economies the urgent need for improvements in logistics in order to ensure an efficient access to the global market. This is particularly important for emerging economies which are away from the principal international trade destinations.

For the specific case of Chile and as opposed to previous studies, this chapter developed a methodology with two sources of added value. First, it is based on the assumption of revealed preferences, avoiding bias through the observation of the real decision process. Second, the validation of the main variables through econometric techniques allows a solid implementation of the K-means algorithm which, in turn, reveals the specific situation in Chile. Previous studies have been supported only by literature review.

Consistent with previous studies, the econometric model revealed that economic and market characteristics, labour market development, port and railway infrastructure are the main variables shaping the potential for logistic zone development in Chile.

At the same time, cluster analysis revealed that the logistics platform in the Chilean market met two differentiated roles: coastal zones supporting port and maritime activities, and inland zones, with a focus on services related to shippers.

At the top of the hierarchy is Santiago, which accommodates 101 ALOG operators whose main focus is on shipper needs. It is followed by two coastal zones, Valparaíso and Concepción, with a strong local market base and very good availability of port terminals and railway. Additionally, this status includes Cordillera, a province whose potential is based mainly on market development.

In relation to public policy issues, the role of port terminals influencing the location decision of logistic operators reveals a successful implementation of Law no. 19.542. The main purpose of this Law was to modernise the port industry through an active participation of the private sector. At the same time, the lack of incentives for railway development would be analysed in order to verify if they are imposing additional logistic cost not only for logistic operators, but also for shippers and even to society through the high externalities associated with motorway development.

Finally, the main shortcomings of the study are related to methodological issues. The first area for improvement concerns the data utilised. The approach does not include cost data, congestion, service time, travel time, land prices, rates and airport infrastructure. Additionally, the utilisation of the number of logistics operators (by the number of ALOG members) as an endogenous variable presents some limitations against other approaches such as business volume or profits, which are not available.

Another important limitation of the study is related to the impact of other provinces on the decisions regarding processes of location. This problem is recognised in the literature as spatial autocorrelation which is treated under the branch of spatial econometrics. Routines to make diagnosis and solve these problems can be found in specialised packages such as GeoDa and MATLAB. Further studies have to take into account the previous shortcomings in order to reach better results.

13.5   References

Arvis, J.F., Mustra, M.A., Ojala, L., Shepherd, B. and Saslavsky, D. 2010. *Connecting to Compete 2010: Trade Logistics in the Global Economy*. The Logistics Performance Index and Its Indicators. World Bank.

Aversa, R., Botter, R.C., Haralambides, H. and Yoshizaki, H.T.Y. 2005. A Mixed-Integer Programming Model on the Location of a Hub Port in the East Coast of South America. *Maritime Economics and Logistics*, 7, 1–18.

Campbell, J.F. 1996. Hub Location and the *p*-Hub Median Problem. *Operations Research*, 44, 923–935.

De Langen, P.W. 2002. Clustering and Performance: The Case of Maritime Clustering in The Netherlands. *Maritime Policy and Management*, 29(3), 209–221.

Doerr, O. and Sánchez, R. 2006. *Indicadores de productividad para la industria portuaria. Aplicación en América Latina y el Caribe*. Serie Recursos Naturales e Infraestructura no. 112. United Nations: ECLAC.

Ducruet, C. 2005. Approche comparée du développement des villes-ports à l'échelle mondiale: problémes théoriques et méthodologiques. *Cahiers Scientifiques du Transport*, 48, 59–79.

Ducruet, C. and Lee, S.W. 2006. Frontline Soldiers of Globalisation: Port–City Evolution and Regional Competition. *GeoJournal,* 67, 107–122.

Logistic Centre. *Definitions for Use.* Europlatforms EEIG. 2004, 16.p.

Hoffmann, J. 2000. The Potential for Hub Ports on the Pacific Coast of South America. *CEPAL Review*, 71, 117–138.

Kara, B.Y. and Tansel, B.C. 2001. The Latest Arrival Hub Location Problem. *Management Science*, 47, 1408–1420.

Leal, E. and Perez, G. 2009. Logistic Platforms: Conceptual Elements and the Role of the Public Sector. *FAL Bulletin*, Issue 274, No. 6. United Nations: ECLAC.

Lu, C.-S. and Yang, C.-C. 2007. An Evaluation of the Investment Environment in International Logistics Zones: A Taiwanese Manufacturer's Perspective. *International Journal of Production Economics*, 107, 279–300.

McCalla, R.J. 1999. Global Change, Local Pain: Intermodal Seaport Terminals and their Service Areas. *Journal of Transport Geography*, 7(4), 247–254.

Notteboom, T. 1997. Concentration and Load Centre Development in the European Container Port System. *Journal of Transport Geography*, 5, 99–115.

Notteboom, T. and Rodrigue, J.-P. 2008. Containerization, Box Logistics and Global Supply Chains: The Integration of Ports and Liner Shipping Networks. *Maritime Economics and Logistics*, 10(1–2), 152–174.

Oum, T.-H. and Park, J.-H. 2004. Multinational Firms' Location Preference for Regional Distribution Centers: Focus on the Northeast Asian Region. *Transportation Research Part E,* 40, 101–121.

Rimiené, K., and Grundey, D. 2007. Logistics Centre Concept through Evolution and Definition. *Engineering Economics*, 4, 89–95.

Rodriguez, V., Álvarez, M.J. and Barcos, L. 2007. Hub Location under Capacity Constraints. *Transportation Research Part E*, 43, 495–505.

Slack, B. 1990. Intermodal Transportation in North America and the Development of Inland Load Centres. *Professional Geographer*, 42, 72–83.

Tongzon, J. 2007. Determinants of Competitiveness in Logistics: Implications for the ASEAN Region. *Maritime Economics and Logistics*, 9(1), 67–83.

van Klink, H.A. and Van Den Berg, G.C. 1998. Gateways and Intermodalism. *Journal of Transport Geography*, 6, 1–9.

Annex 1   Regions, provinces and province capitals

Region	Capital	Province	Region	Capital	Province
I	Iquique	Iquique	VI	Pichilemu	Cardenal Caro
	Arica	Arica		Rancagua	Cachapoal
	Putre	Parinacota		San Fernando	Colchagua
II	Antofagasta	Antofagasta	VII	Curico	Curico
	Calama	El Loa		Talca	Talca
	Tocopilla	Tocopilla		Cauquenes	Cauquenes
III	Chañaral	Chañaral		Linares	Linares
	Copiapo	Copiapo	VIII	Chillan	**Ñuble**
	Vallenar	Huasco		Concepción	Concepción
IV	Coquimbo	El Elqui		Lebu	Arauco
	Ovalle	Limari		Los Angeles	Bío Bío
	Illapel	Choapa	IX	Angol	Malleco
V	Valparaiso	Valparaiso		Temuco	Cautín
	Los Andes	Los Andes	X	Valdivia	Valdivia
	La Ligua	Petorca		Osorno	Osorno
	Quillota	Quillota		Puerto Montt	Llanquihue
	San Antonio	San Antonio		Castro	Chiloé
	San Felipe	San Felipe de Aconcagua		Chaitén	Palena
Metropolitan Region	Colina	Chacabuco	XI	Puerto Aysén	Aysén
	Santiago	Santiago		Coyhaique	Coyhaique
	Puente Alto	Cordillera		Chile Chico	General Carrera
	San Bernardo	Maipo		Cochrane	Capitán Prat
	Melipilla	Melipilla	XII	Puerto Natales	Ultima Esperanza
	Talagante	Talagante		Punta Arenas	Magallanes
				Porvenir	Tierra del Fuego

Index